JEAN-CHARLES TREBBI

promopress

THE ART OF FOLDING

Creative Forms in Design and Architecture

• Jean-Paul Moscovino, Cheval africain, flat and folded.
• The strengthening folds of a hornbeam leaf.
• Santiago Calatrava, architect of the Lyon-Satolas TGV railway station, 1989-1994.
• Crumpled hat made using the techniques of Créations Paula.
• Piramide by Sonia Biacchi, made from sailcloth with flexible fibreglass reinforcements, 2006.
• T.4. cardboard armchair, designed by O. Leblois, 1993.
• Carlo Mollino, Walking Dragons, 1963. Courtesy of Casa Mollino - Turin.
• WASTE.arc: computer-generated model of cardboard housing. Cantercel experimental architecture site, 2008.
• Cover image: detail of a Pietro Seminelli fabric design.

THE FACETS OF FOLDING

ART AND DECORATION

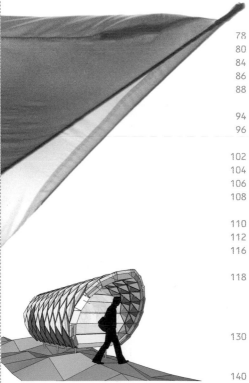

FURNITURE AND ARCHITECTURE

CREASE PATTERNS

The Facets of Folding

"In The Infernal Machine, Jean Cocteau conceptualizes the dimensions of time as a 'folded eternity', imagining a piece of folded fabric that each person pierces with a needle. The needle holes that run through the fabric, markers of a life lived as it passes through the folds of time, are only discovered once it has been unfolded."
Gérard Ayache, Director, L'Institut Infométrie, La Grande Confusion blog, 23 June 2006.

THE ART OF FOLDING

These days, the age-old art of folding goes far beyond dabbling in origami and making paper fortune-tellers and aeroplanes. For many designers, architects and creators, using folds opens up new ways of working with their materials of choice.

I have a long-standing interest in this area. As a child I was always proud of showing off my creations, which included a paper aeroplane designed to look like a twin-engine plane . My fascination has continued into the present day, whether through creating representational origami or through discovering new shapes in geometric vectors. I experienced equally magical feelings as I discovered the architectural designs, decorative objects and pieces of contemporary furniture that are presented in this book.

My initial involvement in this area was based principally on creating forms, whether from a sheet of paper or from other materials. The idea was to be able to create an object with the minimum possible loss of material. Amongst the many marvels that I discovered through the lectures of Raymond Guidot, an engineer, design historian and lecturer at the Ecole Nationale Supérieure des Arts Décoratifs, my eyes were opened to the furniture pieces of, Alvar Aalto, Charlotte Perriand, Carlo Mollino and many others besides.

Then came research into the simple mechanisms of children's pop-up books, which provide movement in 3D whilst always adopting a minimalist approach. Then the discovery of Richard Sweeney's superb curved sculptures was complemented by my curiosity-fuelled foray into Eric Gjerde's magnificent tessellating creations, as well as the fractals and twists of Chris Palmer and Jeremy Shafer, and the deployable structures of Sergio Pellegrino, Simon Guest and Davide de Focatiis. So far an inspiring research experience!

Delving deeper into Masahiro Chatani's *Origamic Architecture* drove me to create some pieces of pop up art and architectural origami of my own, and also to experiment with creating light and shadow effects using paper. Studying landscape architecture and cave dwellings gave me inspiration for my 'trogligami' pieces. Then, looking into the aesthetics of the folds in organic forms, the works of Professor David Huffman, a pioneer of mixing art with computer science, and his research into curvilinear folds brought me closer to designing illuminated sculptures. I had always loved light effects and creating clarity and opacity through working with pliable materials or paper of various textures.

Another area I became passionate about was crumpling. This technique is one of the subjects of study led by Vincent Floderer at the Centre de Recherche International de Modélisation par le Pli (CRIMP). His works, inspired by those of Paul Jackson, showed me the relationship between nature and folding, which interested me from the perspectives of the natural and geometric shapes of bionics and biomimicry. This opened up yet another field to explore! A basic radial pattern and successive folds, along with Romain

- *Mr Métro, designed and folded by Claudine Pisasale.*
- *Origami made by Max Hulme at an MFPP get-together in Paris, May 2008*
- *Modular, multi-sided light, designed and folded by Claudine Pisasale*

Chevrier's tubular crumpling techniques, brought surprising results! And although crumpling, which superficially appears to be a multitude of unorganized folds, did not initially seem to belong to the world of folding in my view, an encounter with the internationally-known and indefatigable origamist Claudine Pisasale, which led me to discover the superb fashion creations of "Paula" by Annette and Paul Hassenforder, prompted me to change my views entirely.

Great contemporary international experts in the field of folded art such as Robert J. Lang, Paul Jackson, Eric Joisel, Korio Miura (the Japanese astrophysicist and inventor of the Miura fold), Tomoko Fuse, Eric Gjerde, Goran Konjevod, Sumiko and Yoshihide Momotani, Jun Maekawa and Max Hulme all endeavour to disseminate and share their knowledge. Clubs such as the Mouvement Français des Plieurs de Papier (MFPP), created by the artist Jean-Claude Correia in 1978, hold international conferences at which knowledge and know-how are shared. Nevertheless, many artists and designers remain unaware or on the side-lines of these trends

With just an ordinary sheet of paper and a craft knife, and by observing and copying natural shapes, or through random experimentation with geometric folds, a fascinating world reveals itself. Whether you imitate natural forms or consciously avoid them all comes down to the heat of the moment. As Nicole Charneau says, "If you give a child a sheet of paper when he or she doesn't have any coloured pencils to play with, his or her first reaction will be to fold it. That's the spontaneity of folding! Just as the folds of the waves animate the sea, a single fold in a sheet will bring any flat surface to life."

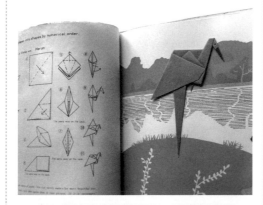

• *Happy Origami book series from the 1960s - Turtle Miyawaki, Tatsuo Biken-sha, 1964. Claudine Pisasale's personal collection.*
• *Traditional origami swan made using modular origami, by Claudine Pisasale.*
• *Origami chess set made by Max Hulme at an MFPP get-together in Paris, May 2008.*
• *Origami figurines made from metro tickets by Lee Shang.*

Cette boîte à bonbons, de forme si originale, peut être construite facilement par le pliage de 2 carrés de papier, et sans une goutte de colle.

Pour bien en comprendre le pliage, je vous conseille de la faire d'abord avec une feuille de papier blanc ordinaire, puis vous pourrez la fabriquer avec du papier de couleur un peu fort (papier parchemin glacé).

Prenons un carré de papier de 20 centimètres de côté ABCD, et divisons-le en 16 carrés égaux de la manière suivante :

Mettons la ligne C D sur A B et faisons le pli E F. Remettons à plat. Mettons A B sur E F et plions suivant I J, C D sur B F, et plions suivant K L, et remettons à plat tout le carré (fig. 1).

Mettons ensuite A C sur B D, et plions suivant G H, remettons à plat ; mettons A C sur G H et plions suivant M N, puis B D sur G H et plions suivant O P. Remettons à plat tout le carré, et

nous voyons qu'il est divisé en 16 carrés de 5 centimètres de côté. (fig. 1). Plions maintenant les quatre coins I A M, J B O, L D P et K C N (fig. 2), et coupons-les suivant les lignes M I, O J, L P et K N. Cela fait, entaillons avec des ciseaux, suivant les 4 lignes en traits forts G Q, F T, H R et enfin H S (fig. 2).

Plions suivant les lignes G F, F H, H B et E G (fig. 3) et nous obtenons un carré qui est la moitié du carré primitif.

Les entailles vont nous permettre maintenant de faire 4 nouveaux plis, après chacun desquels nous remettrons le papier à plat dans la position de la fig. 3. C'est d'abord le pli R T (fig. 4), obtenu en mettant F H on F' H'. Faites de même les plis Q T, Q S et S R (fig. 4).

Une fois le papier remis à plat dans la position de la figure 3, nous voyons que nous avons formé les triangles e, f, g, h, ainsi que les 8 triangles a (a) d (d) c (c) et b (b).

Les plis U V, V Y, Y X et X U de la figure 5 sont des plis en relief ; il faut les transformer en plis creux en amenant au centre les pointes G, F, H et B. Nous formerons ainsi la figure 5, en laissant les 4 plis U V, V Y, Y X et X U en creux.

Remettons alors le papier dans la position de la figure 5, et nous voilà prêts à monter la boîte.

Le fond de cette boîte est le triangle S Q T R (fig. 5).

Les flancs se composent des 4 triangles e f g h et des 4 triangles a b c d, soit 8 triangles en tout.

Comme il y a 12 triangles sur la figure 5, on voit que 4 doivent disparaître, mais il ne s'agit pas de les couper ; ce seront, au contraire, ces mêmes plis qui nous seront utiles pour la construction.

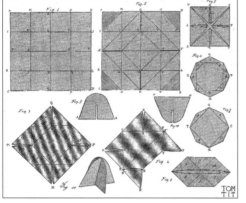

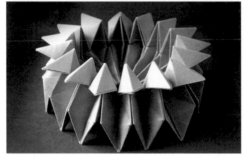

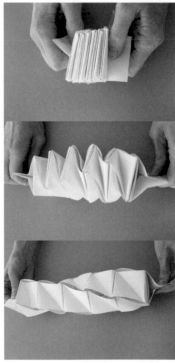

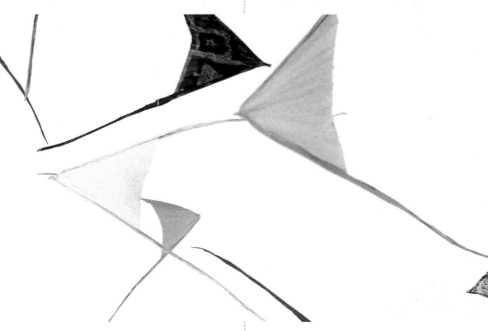

• Joujoux en papier par Tom Tit,
*Paul Lechevalier éditeur, 1924, from
Jean-Jérôme Casalonga's personal collection.*
• *Twisting, stretching origami by Claudine
Pisasale.*
• *Circular paper fortune-teller, designed and
folded by Claudine Pisasale.*
• *Crumpled, coloured fish, designed by Paul
Hassenforder, folded by Taki.*

THE WORLD OF FOLDING

The word origami, which comes from the Japanese 'oru', meaning 'fold', and 'kami', meaning 'paper', literally denotes the actions of the art form. More than simply being a technique, it is above all a way of thinking differently about how to give form to an object.

The classic way of teaching origami is through the master providing the student with verbal instructions and through observation and repetition. According to some researchers this ancient technique was born in Japan, while others argue it originates from China. The first book on origami, Hiden Senbazuru Orikata ('The Secret of Folding One Thousand Cranes'), was published at the end of the eighteenth century. The book, authored by the Buddhist monk Rokoan, provides a selection of folded crane models. This bird, a symbol of happiness and immortality, encapsulates the Japanese aesthetic. The book contained the first ever sets of detailed instructions for making origami, or 'crease patterns' as they are known in today's terminology.

With its mixture of magic and instinct, origami often seems like a sleight of hand. In many respects it's a clever game that involves selecting the right paper – rough, smooth, glossy, satin or silky – knowing the right techniques – folding in half, gluing or wetting the paper – and of course getting the folds right. Origami, or the art of folding, became a fully-fledged science some time ago, all the while retaining its magic and poetry.

Origami made the transition from traditional to modern with the help of Shuzo Fujimioto, the father of the geometric fold. But we have to thank the origami master Akira Yoshizawa for providing us with a vocabulary of signs and symbols – the origamist's equivalent of solfège – that allow us to differentiate valley folds from mountain folds, an advance that has allowed origami to break out of the restrictions of language barriers and be understood visually.

For a long time and for obvious reasons, the art of folding was limited to paper, textiles and other materials that can easily be worked with naturally. But current trends – and this is precisely the purpose of this book – have broken away from these restrictions. The emphasis is now on finding new shapes and forms, sometimes with the help of computer technology, which still start out as a flat piece of material, but are not restricted by the chosen material. For now practically any material, concrete included, can be 'folded'. Much of the progress made has drawn on nature. One only needs to look at the contours of the land, the texture of cabbage leaves, the delicate pleats of palm leaves and the intricacies of animal skins to see that nature is certainly not sparing in its use of folds. This imitation of nature, known as 'biomimicry', has been the subject of some important pieces of research. Dennis Dollens, an architect based in New Mexico, has been studying the structural forms found in nature as part of his research into sustainable development. In his view great innovations can come from observing living things and their structures and mechanisms. The work of Dr Taketoshi Nojima of the University of Tokyo is based around fold patterns found in nature, and in particular the growth processes of plants. It's a whole new scientific field, very accurately christened 'BiOrigami' by Jean-Jacques Dupa, an engineer and researcher at the Commissariat à l'Energie Atomique in France.

For many years the furniture and décor communities were more interested in the 'foldable' than the 'folded'. Through various innovations that made use of bolts, hinges, pivots and rotational systems, designers put forward some ingenious piece of work that are more compact and easier to store. Before becoming interested in the 'folded', architects made use of flexible and deployable designs, in the form of mobile structures that range from tents to the Mongolian yurt, and even to Guy Rottier's cardboard house, built in 1968, or the kinetic architecture of Chuck Hoberman. But today the notion of folding is not restricted to the merely 'foldable' and responds to imperatives that go beyond compactness, revealing an infinitely rich range of applications.

It's easy to find a prodigious range of objects all around us that are either foldable, folded, pleated or deployable. These objects can be magical, fun, practical or simply pleasing to the eye; and they can be based around an ingenious use of folding or mathematical prowess. The art of folding, a passion for thousands of people across the world, will surely continue to surprise us.
Enjoy this guide to it!

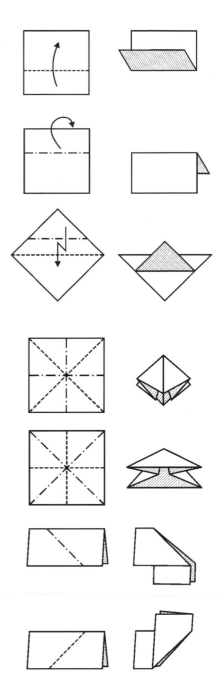

The solfège of origami according to international symbols and the MFPP. From top to bottom:
- *valley fold.*
- *mountain.*
- *pleat fold.*
- *square base.*
- *water bomb base.*
- *inside reverse fold.*
- *outside reverse fold.*

✉ www.lagrandeconfusion.com/?Hypermonde;
www.miura-ori.com/English/e-index.html;
www.joanmichaels.paque.com/;
www.richardsweeney.co.uk/; Ray Schamp:
http://fold.oclock.am/; http://mfpp.free.fr/; Jean-Claude
Correia, www.magnol.com/jm/ correia.htm; Optigami
and TreeMaker by Robert J. Lang; Junior Fritz
Jacquet: http://www.leplieur.
com/, Matthew Shlian: www.mattshlian.com,
www.guidodaniele.com.

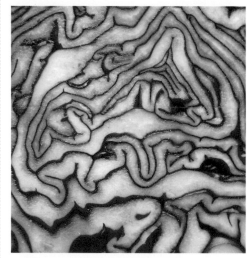

• Guido Daniele, Elefante - Campagna Europea
Schroeders - 2002. A multimedia artist, since
2000 Daniele has been making an extraordinary
hand-art series called Mani Animali using body
painting techniques.
• Comparison of the folds of a crumpled origami
mushroom by Vincent Floderer and some Antilles
coral.
• Convolutions of a red cabbage.
• The natural folds found in the leaves of
chamaerops humilis var. fourasiena
(Mediterranean dwarf palm).

In addition to being a psychiatrist and psychoanalyst,
as well as holding a doctorate in psychology, Serge
Tisseron is also a writer and designer. In his book
Petites mythologies d'aujourd'hui (Editions Aubier,
2000), he describes with great accuracy the senso-
rial nature of the fold. "Proust mentions these
Japanese paper pellets which, when thrown into
water, unfurl themselves to become animals or flowers. Origami is another manifestation of this art of
folding. In both cases the wonderment of the spectator comes from discovering that a three-dimensional
form can emerge from a single surface area. In the
case of origami, paper ceases to be for packaging
three-dimensional objects such as sweets or
presents, and instead packages a virtual space. No
shape is imposing its form or features on the paper.
As fold follows fold, some become partly hidden, only
to become visible once more, only now transformed
by the other folds and made impossible to distinguish
because of the new positions taken up by new sides
created in the sheet. In origami, folds only hide more
folds. The surface of the paper only shrouds its own
transformations. Bachelard argued that our
culture, and most notably the field of psychoanalysis,
only takes formal imagination into consideration,
forgetting dynamic imagination and material
imagination. With paper, such reductionist errors are
impossible. Paper has no shape, yet it can take any.
And this pliability is perhaps the main reason for the
fascination that folding inspires. Paper is a material
that, in being twisted again and again, creates totally
unexpected forms that owe nothing to the paper's
original surface area and everything to the forces
that have been applied to the paper. A folded piece
of paper has its own topology, the characteristics of
which can only be explained by the forces that have
been applied to it. And these forces can in turn only
create shapes if the sheet of paper lets them be
imposed on it without tearing, or in other words

without stopping to exist as a single and continuous surface. But isn't this the best way to represent the way we function from a psychological standpoint? Within the layer of our perceptions, our psychological faculties fold and re-fold our experiences to create representations, models and forms."

AEROGAMI

Thousands of paper plane fanatics enjoy this economical, quiet and environmentally friendly hobby. The last world paper plane championship took place in the May 2006 in the legendary Hangar 7 of the aeronautical museum in Salzburg, Austria. Forty-seven different nationalities competed across three categories: longest duration flight, longest distance flight and aerobatics. But the competition winners did not break the records held by the American stars of the field. The aeronautical engineer Ken Blackburn still holds the longest duration record for the 27.6-second long flight he managed in Atlanta in 1998, while Tony Fletch broke the record for longest length of flight in 1985 with a 58.8 metre effort.

Tests carried out by MFPP members Olivier Viet and Yves Clavel with the expert collaboration of Amar Tarabit, President of Aéro-Club Paris Sud, demonstrate the complex principles of flight and the forces that are exerted upon the humble folded sheet of paper. Here are some of their tips for a successful flight. Firstly, to ensure that the plane stays steady, it is important that the angle formed by its two wings is less than 180 degrees. Ailerons keep the craft stable through being at a slight angle, which makes the plane turn gently to the left or the right. To do a loop-the-loop, it is necessary to launch the plane with the forces of lift and gravity that are exerted on the plane as air flows over its wings.

🖰 www.paperplane.org; Red Bull Paperwings; Paperang by Edmond Hui; Didier Boursin; boomerang plane by Norio Torimoto; Avenger paper plane made by students from the University of Leeds; origami planes by Olivier Viet; www1.atword.jp/pavionaparis/; http://oriplane.com/; Sciences et Avenir magazine, August 2002; http://www.guidodaniele.com/.

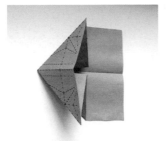

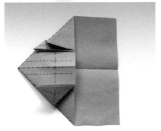

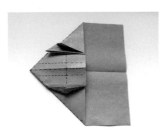

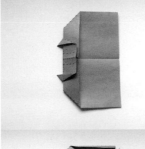

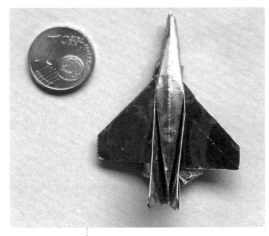

• Folding steps for the Papion plane (see also the diagram at the end of the book).
• In Japan, aeroplane origami is called kami hikôki, from the words kami (paper) and hikôki (aeroplane). Toda Takuo, President of the Japanese Origami Aeroplane Association, came up with the idea of constructing a 26m-high, purpose built paper aeroplane launch tower; it is the only one in the world. In 2001, Takuo founded the Paper Aeroplane Institute in Fukuyama, Hiroshima Prefecture. A world championship is due to be held there in 2010.
• Miniature aeroplane, designed and folded by Yoshidide Momotani at an MFPP get-together in Paris, May 2008.

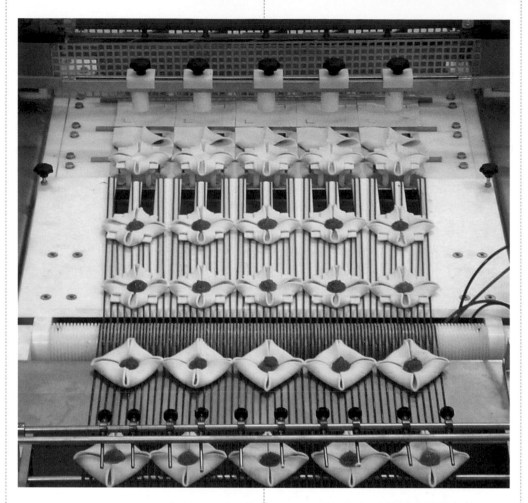

ORIGAMI FOR FOOD-LOVERS

How could one possibly talk about the art of folding without mentioning folded foodstuffs such as croissants, samosas, Lyon fritters, Danish pastries and Bolognese tortellini (which legend has it is based on the shape of Venus's navel)? And let's not forget about the Moroccan delicacies of chebakai (a biscuit) and meloui (a pancake), or Mexican burritos. A gourmand folding enthusiast is never short of food to satisfy his cravings! What is less well-known is the use of plants in making folded art. In certain countries, plant leaves that are usually associated with weaving techniques, such as palm leaf and banana leaf, are not only used as wrapping or to make receptacles for food, but are also used, along with coconut palm, to astonishing effect in making origami animals (grasshoppers) and people. Design company Escarboucle sells a range of skilfully crafted dishes made from moulded palm leaves. The dishes are made in India from this natural fibre. They are waterproof and 100% compostable, and they can withstand water temperatures of up to 100°, oil temperatures of up to 150°, 220° temperatures in the microwave and -25° in the freezer.

✏ **Danish pastries; www.formfrys.dk;
Baï Tong in Thailand; www.philcad.com;
www.escarboucle.com.**

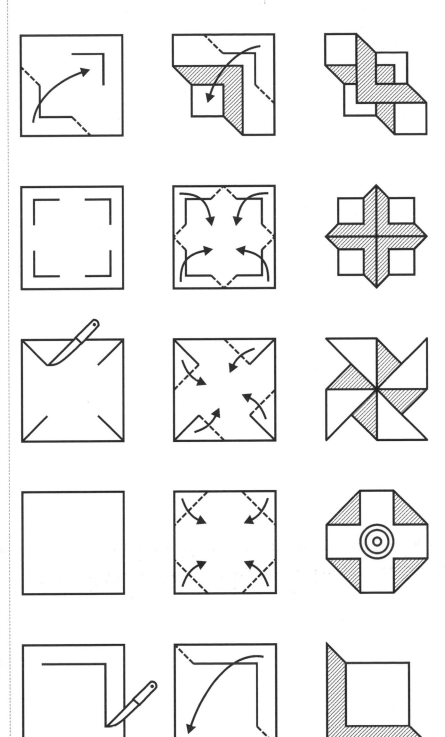

• Making pastries
Here are a few ideas from philcad.com for some folds and cuts for use when making classic pastries. After rolling the pastry to a thickness of 2.8mm, cut it into squares with a width of 9.5cm, then fold as shown in the diagrams. Place in the middle of a tray of your choice. Brush with egg wash and bake in a hot oven (220°) for approximately 20 minutes.
Windmill: Make slits in the corners of the square and fold at an angle.
Viennese square: Fold the four corners of the square towards the middle and pinch the edges.
German square 2: Make right-angled cuts at two diagonally opposite corners and fold as shown in the drawing.
An alternative version of the German Square 3: Make right-angled cuts at the four corners and follow the instructions in the diagram.

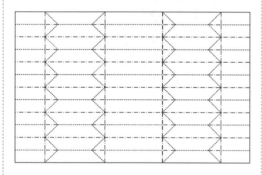

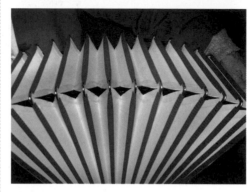

• *Different forms that can be produced from a basic accordion fold.*
• *Crafting bellows requires a meticulous and precise approach. In these preparation stages, you can see the pleated card strips before they are cut, and also the coloured (in this case blue) exterior that follows the inside of the bellows.*
• *Le papier du Père Mathieu, models from accordion folds, print from Joujoux en papier par Tom Tit, Paul Lechevalier éditeur, 1924, from Jean-Jérôme Casalonga's personal collection.*
• *Carlo Mollino, Walking dragons, 1963. Courtesy of Casa Mollino - Turin.*
• *Right-hand page: Carlo Mollino, Del drago da passeggio greetings card, 1963. Courtesy of Casa Mollino - Turin.*

ACCORDION BELLOWS

Contrary to what you might have thought, the bellows of an accordion are not made from a single, folded sheet, but rather from four panels. "An accordion works through air being pushed through the reeds, which makes them vibrate. The bellows are integral to the instrument's capacity for musical expression. The bellows are made out of a special type of card known as 'leather board' or 'carte de Lyon'; four panels of the card are folded and bound together with goatskin. The cardboard is covered with satin and the folds of the bellows are reinforced at the corners with stainless steel. The two extremities of the bellows are attached to a wood frame, which must fit the body of the accordion perfectly."

✎ **Accordéons de France – Maugein Frères - 19000 Tulle - (France)**

WALKING DRAGON

"The folding technique used in this toy is the type used in camera bellows. The dragon has a cylinder hidden under its head that was held in place on each side by an elastic band which, when wound up, causes the cylinder turn and made the dragon move, an age-old system used in certain types of toy. The dragon was a 1963-1964 New Year's gift that the genius architect and designer Carlo Mollino gave to a dozen of his closest friends. He had bought the bellows from La Rinsascente department store and decorated them by hand, each one differently, and also signed and gave a name to each dragon. The dragons were placed folded up in clear plastic boxes with holes in the lid (to let the dragon breathe!) and were accompanied by a humorous booklet and two photographs taken and colourized by himself."

✎ **Exhibition at the Chambre de commerce de Paris, November 2007. Remarks by Napoleone Ferrari, Courtesy Museo Casa Mollino – Turin**

TROUBLEWIT

To confound their audiences, conjurers and magicians have often performed tricks using a long strip of paper that is pleated like an accordion, known as Troublewit. With just one piece of paper and sufficient dexterity and a very good memory, it is possible to make between 60 and 70 different shapes. The trick, which has been in existence since at least 1710 – and possibly even long before that – was often used by Harry Houdini.

TABARIN'S HAT

This magic trick was invented by the French entertainer Tabarin (1584-1633). It consists of a simple, black felt disc which through a variety of twists, folds and other manoeuvres is transformed into many different shapes and forms. Depending on the story being recounted, the entertainer in his show can variously don a pirate hat, an Oxford fellow's mortar board, a Tyrolean hat, or the headgear of a samurai or clown. An illustration published in the September 1933 issue of Popular Science provides us with some of the hat's amusing fold patterns.

HAT TRICKS

In his workshop-cum-show, Paul-Henri Jeannel creates magic hats using the art of paper folding. Both a teacher and a magician, this virtuoso of paper shows you how – and without using either glue or scissors – to quickly make fun and original headgear such as a princess's tiara or a pixie hat.

• 13 x 18 cm tourist camera, in production between 1890 and 1930. From Lionel Turban's collection.
• Paul-Henri Jeannel creating one of his magic hats.
• Magic Hat, Popular Sciences, September 1933.

☞ "Origastelet" by Justin Lenoir: **http://theaterruepietonne.free.fr/Theater-Rue-Pietonne_Spectacles3.html** ; Pierre Guedin: **arte-fake.com**; Arturo Brachetti; **www.funandibule.com/irigimi**; Kaminokami; **http://chapeau-magique.net**; Jeremy Shafer: **www.barf.cc/**; Patrice Curt, Magica Planet; "origamagique" by Ludovic Toulouse; "Orikadabra" street theatre by Marieke de Hoop; "Irigimi" by Julien Gritte.
☞ **Lionel Turban: www.disactis.com**

VIEW CAMERAS

Made from a single piece of material, the bellows of a view camera are different to those of an accordion. Lionel Turban is a camera enthusiast and a specialist in artisanal and film photographic and cinematographic techniques, and also in nineteenth century photographic processes. He has reconstructed several different pieces, including one of Louis Lumière's cameras, which is on display at the Musée du cinéma et de la photographie in Saint-Nicolas-de-Port. At his website you can find instructions for making your own view camera, not to mention a wealth of information and tricks related to the photographic processes of the nineteenth century.

STENTS

We can find an example of origami folding techniques being used in the world of medicine in the form of stents. Stents are very tightly folded tubes that are frequently used in certain types of operation to prevent arteries from collapsing. Dr Zhong You and Kaori Kuribayashi of the University of Oxford designed some ingenious expandable stents that can stretch from 12mm to 23mm in diameter.

CAPILLARY ORIGAMI

Benoît Roman, Charlotte Py, José Bico, Charles Baroud and their colleagues at the Ecole Polytechnique and the Ecole Supérieure de Physique et de Chimie Industrielles (ESPCI) in Paris are practitioners of 'capillary origami', a complex field that presents exciting opportunities. To put it in simple terms, this type of origami is produced through depositing a drop of water on a very fine elastic membrane. When it evaporates, the capillary forces cause the membrane to bend. These clever researchers believe that "capillary origami is relevant for developing three-dimensional microstructures from two-dimensional patterns. When working at a small scale capillary forces are very powerful, meaning that little drops of water could effectively be used like micro-pincers or tweezers." And of course other applications are yet to be discovered!

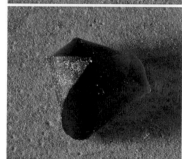

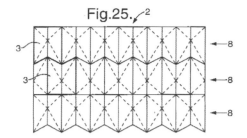

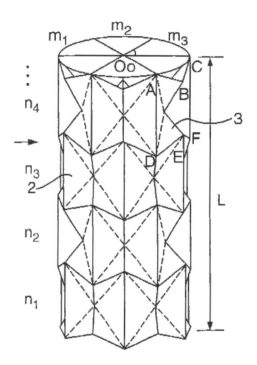

• Capillary origami pyramid made from a single droplet of water.
• Diagram of the folds of a stent.

TESSELLATIONS

One field that is rapidly developing is the use of geometric shapes such as tessellations. This origami technique makes use of paper modules to build spectacular compositions. The most important advances in this area were made during the 1960s and 1970s by the origami masters Shuzo Fujimoto and Yoshihide Momotani, and have since been developed into new styles by Paulo Barreto, Chris Palmer and Joel Cooper.

Eric Gjerde, a keen user of geometric shapes and tessellations in his origami, wanted to be a 'paperologist' from the age of five. A key influence in developing this style, Eric takes great pleasure in sharing his creations with the world through the tutorials and showcases of his work on his website (www.origamitessellations.com). He is always eager to try new materials, and works masterfully with his tessellations to give them real volume.

Over in Switzerland, Mélisandre creates magnificent and complex folded works that draw their inspiration from the tessellations found in such places as the churches of Rome, the domes of a mosque in Toledo, or the flowers in her garden. She's a self-effacing person who likes to share: "My art is not the product of some genius working alone in some ivory tower; it is nurtured through exchanges with artists from across the world. A diagram for an origami model has the same function as a musical score, namely to let another artist produce their own interpretation of the work".

✎ **Arthur Stone, Flexagons and flexatube; Harold McIntosh, trigonal and hexagonal flexagons; Martin Gardner; Bud by Jeannine Mosely; Jun Maekawa; Joseph Wu; modular origami by Tomoko Fuse; Pliajeux by Thierry Chapeau; Eric Demaine; www.origamitessellations.com; Dr David Huffman, pioneer of research into curved folding in computational origami; Origami by Mélisande: http://origami-art.org/blog/; "Tess" by Alex Bateman and "Tree Maker" by Robert Lang.**

DANCE AND ORIGAMI

The Italian origami-dance concept show "Gigantorigamico", developed by Sonia Biacchi, who in 1982 founded the CTR (Centro Teatrale di Ricerca), and featuring the origami creations of Luisa Canovi, was the first fusion of music, dance and origami seen in Europe. Fusing these artistic forms created a magical and unique spectacular. In Australia and the United States, a group called BalletLab produced a show that combined the pliés of ballet with the folds of origami. The visual impact of the movements associated with folding origami helped to create unique choreographic sequences. In France, Véronique Wardega collaborated with Compagnie Chantier Majeur on a choreographic project that allowed her to explore the world of folding art on a larger scale, working with squares that were up to 3 metres wide! Performed at the Montgeron theatre in December 2004, the show explored folds both large and small, mixing origami, music and contemporary dance. Collaborating with Stéphanie Roussel, Frédérique Robert and Christophe Tessier, the team created playful and moving wild cranes for the event.

✎ **www.ctrteatro.com; BalletLab: http://www.abc.net.au/rn/deepend/stories/2006/1687151.htm**

• *TGV, origami tessellation by Christiane Bettens, made during a TGV train journey to Paris in 2007. The piece measures 20 x 24cm; the paper from which it is made was roughly double those dimensions. Made from Muresco paper, an Argentinian-produced hard-wearing silk paper with a satin finish on one side. Based on a 3 x 64 grid.*

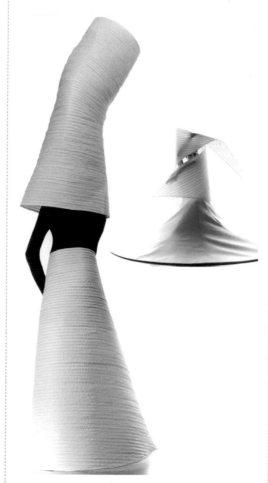

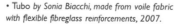

• Tubo *by Sonia Biacchi, made from voile fabric with flexible fibreglass reinforcements, 2007.*
• Campanula *by Sonia Biacchi, 2004. The skirt is synthetic leather, while the upper half of the costume is made from sailcloth with fibreglass reinforcements.*
• *Facing page:* Colomba *by Sonia Biacchi, 2000. Skirt made from laminated wood and stainless steel.*

• Ribbed industrial cladding.
• Jean-Pierre Campredon, architect, used cold
formed polycarbonate to construct the
transparent folded exterior of the Cantercel
architectural experimentation site.
• Glass façade of the Citroën Immeuble C42
building, Champs-Élysées, Paris, designed by
Manuelle Gautrand, architect 2007.
Right-hand page:
• Santiago Calatrava's Lyon-Satolas TGV railway
station (1989-1994).
• Foreign Office Architects, F.O.A, Yokohama
International Port Terminal, Yokohama Japan,
2002.

ARCHITECTURE

With the balancing acts it demands in relation to materials, weights, stress, loads and other forces, of all the fields of creative endeavour architecture is perhaps one of the most complex and most deserving of its own dedicated study. For some designers, folding is a language and a lexicon of its own. Having been the subject of research and experiments in the 1950s and 1960s, the structural efficiency of folded structures has been proven beyond doubt. The architect Patrick Bouchain provides us with a clear description of this phenomenon: "If you rest a sheet of paper between two supports, it bends under its own weight because it has no inertia and therefore no rigidity. But if we create a series of parallel folds that follow the direction of the space that the paper spans, it becomes rigid. You see the same principle at work in folded framework; the idea is to significantly increase rigidity. There are many different types of folded structures, including structures that make use of bays, radial structures and arch or portico features. The underlying principle is that of a module that is repeated or transformed by rotation (radiating from the centre), translation, duplication or overlaying".

Creative and visionary men and women have dedicated their lives to developing techniques and methods that revolutionize traditional know-how. Jean Prouvé (1901-1984) is one such man.
He popularized a folding machine, changing the way that furniture and architecture are produced. Thanks to the machine, metal went from being a fairly unwieldy material to one that could finally be produced in folded sheets and therefore in sections. This opened up a new world that brought in new production methods and working patterns.
Pre-lacquered sheet metal (made variously from galvanized steel, stainless steel, lacquered aluminium, copper or zinc) can be found in many industrial sectors. It has numerous applications in the construction industry, including in façades and protective cladding, formwork and suspended ceilings.

By the 1980s Frank Gehry and Peter Eisenman had ushered in new deconstructivist tendencies, which incorporated folds and multiple faces. Notions of roofs and walls disappeared, making way for skin-and shell-like structures of various textures that let architects' imaginations run riot.

One of the greatest architects and engineers of our time is undoubtedly Santiago Calatrava, whose work combines great technical expertise with elaborate organic shapes. Amongst his many works, the Ernsting Warehouse in Coesfeld (Germany, 1983-1985) is unique for its set of folding doors. An even more sophisticated design of his is the Lyon-Satolas TGV railway station (1989-1994), which is recognizable through the soaring majesty of its outspread wings.

Sophia Vyzoviti, an architect and lecturer at the Delft University of Technology in the Netherlands, has been experimenting with new folding techniques have since 2001, with the goal of intuitively discovering new structural forms. The fruits of this research can be used in a range of architectural sub-fields, as well as in the fields of fashion and product design.

✏ www.bardage-metallique-perfore.tolartois.com /; http://www.metaldeploye.com/; Industrial Origami; Wuko Pli mini manual folding machine, Philippe Potié, "Autour de la plieuse de Jean Prouvé". Imaginaire technique, Parenthèses, 1997.
Luigi Moretti; Massimiliano Fuksas, tower house Frankfurt am Main; EPFL, Yves Weinand, wooden pleated structures, Félix Candela; Tim Tyler: http://pleatedstructures.com/; http://www.dillerscofidio. com/; Patrick Bouchain, structures: http://www.crit.archi.fr/ Web%20Folder/bois/; Mac Farlane; Grimshaw; Peter Eisenman, BFL Limited Bangalore 1996; FOA; www.digit-all.net; Etienne Cliquet, www.ordigami.net/; Tadao Ando ; Pascal Amphoux: "Runninghami" noise barrier ; www.paramodern.com; www.calatrava.com/; Shuhei Endo, Springtecture: www.cantercel.com/informations/; parallel fashion and architecture MoMa Los Angeles, "Skin + Bones" exhibition 2006. Show room Issey Miyake, "Pleats please", Berlin 2004, by Ammar Eloueini.
H2 house by Greg Lynn, 1996

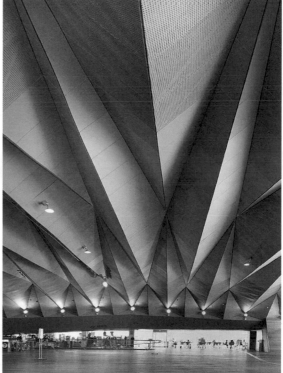

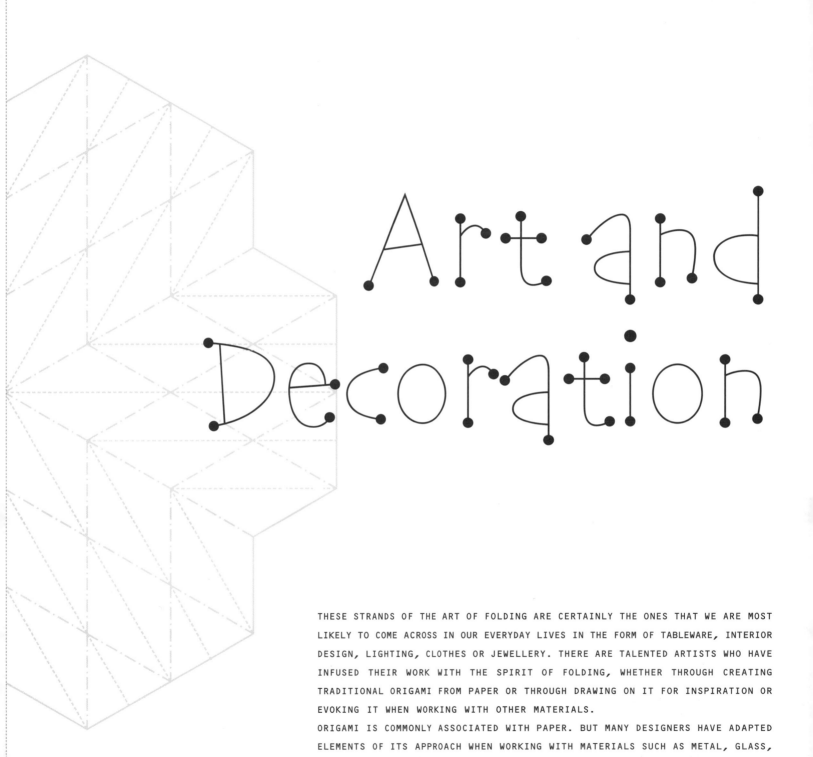

Art and Decoration

THESE STRANDS OF THE ART OF FOLDING ARE CERTAINLY THE ONES THAT WE ARE MOST LIKELY TO COME ACROSS IN OUR EVERYDAY LIVES IN THE FORM OF TABLEWARE, INTERIOR DESIGN, LIGHTING, CLOTHES OR JEWELLERY. THERE ARE TALENTED ARTISTS WHO HAVE INFUSED THEIR WORK WITH THE SPIRIT OF FOLDING, WHETHER THROUGH CREATING TRADITIONAL ORIGAMI FROM PAPER OR THROUGH DRAWING ON IT FOR INSPIRATION OR EVOKING IT WHEN WORKING WITH OTHER MATERIALS.

ORIGAMI IS COMMONLY ASSOCIATED WITH PAPER. BUT MANY DESIGNERS HAVE ADAPTED ELEMENTS OF ITS APPROACH WHEN WORKING WITH MATERIALS SUCH AS METAL, GLASS, CERAMICS OR COMPOSITES. EXPERIMENTING WITH FOLDING IN THIS WAY HAS YIELDED INTRIGUING RESULTS.

• *Kyouei Design's,* Honeycomb lamp, *made from denguri paper, 2007.*

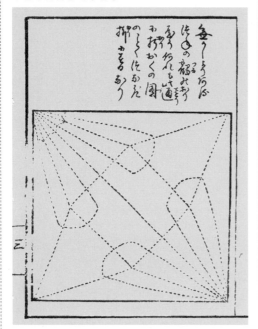

• A page from one of the first origami hobby books, Hiden Senbazuru Orikata (1797).

The Folded Crane: Orizuru

"With its simple beauty, the crane is the piece of origami that best represents the Japanese aesthetic. Symbolically speaking, the crane and the number one thousand are closely linked. They say it is the bird that lives for a thousand years; the origami crane goes back to the year 1000 (one thousand being the divine figure of happiness and immortality). In Japan, the elderly are often given gifts of paintings or prints that depict cranes, tortoises or pines. Origami cranes are also given as gifts to the sick, a gesture that means "we wish for you to get better and live for a thousand years". Although tradition dates the origami crane back to the year 1000, no document prior to 1682 attests to its existence, with the first illustrations of it appearing in 1700. From this period onwards the motif of the origami crane has often been used in textiles, and prints frequently depict young girls and geishas folding cranes. In 1797 Hiden Senbazuru Orikata (The Secret of Folding one Thousand Cranes) was published, and is regarded as the first origami hobby book. First Published in Kyoto by Tamehachi Yoshinoya, it contains forty-nine crane models, each one accompanied by a humorous kyoka poem. The author of the fold designs was the Buddhist monk Rokoan; the kyoka poems were by Rito Akisato and the illustrations by Shunsenai Takehara. The first ever crease pattern appeared in this work.

Origami cranes became famous across the world after World War Two. Amongst the victims of Hiroshima and Nagasaki was a little girl named Sadako Sasaki, who was two years old when the bombs were dropped. Ten years later she fell ill with leukaemia, which in Japan was known as 'bomb sickness'. Having heard that her wishes would come true by folding 1000 paper cranes, Sadako set about doing just that so that she would get better. She managed 644 before she died, but her classmates folded the remaining 356, and also published the letters that Sadako had written to them from her hospital bed. The letters went around Japan, and the story of Sadako and the thousand cranes became famous. Her friends had a monument put up in her memory at the Hiroshima Peace Memorial Park. Sadako stands on top of a dome, folding the wings of a crane. At the foot of the monument, children from across the world present garlands of a thousand cranes as an offering for world peace."

RECOLLECTIONS OF MEETING WITH TAKI GIRARD AT AN EXHIBITION AT THE CENTRE CULTUREL DU MOULIN, LONGVIC, DECEMBER 2007.

• Cranes based on a traditional design, made using a crumpling technique. By Taki.

TAKI

A visual artist, Taki makes creations that are full of finesse and construct a delicate and highly poetic world. Her favourite folds and creases stem from the symbolic figure of the crane. She also creates children's books that contain poems that are accompanied by origami pieces, which can be hidden under his or her pillow. Origami is a state of mind, and when she is folding she savours feeling her hands working in perfect harmony.

Origami creations by Taki.
• One-thousand crane composition.
• Poetry book for children.

PAPER: A COMPLEX MATERIAL

ON THE FACE OF IT, PAPER WOULD APPEAR TO BE THE SIMPLEST MATERIAL TO FOLD. BUT DIFFERENT PAPERS ARE MADE OF DIFFERENT MATERIALS AND VARY IN TERMS OF HOW THEY RESPOND TO BEING FOLDED, RANGING FROM PHOTOCOPIER PAPER THROUGH TO THE RENOWNED "WASHI", A TYPE OF PAPER FROM JAPAN MADE BY HAND USING "KOZO", "GAMPI", "MITSUMATA" PLANT FIBRES, WHICH ARE LONGER THAN STANDARD WOOD FIBRES.

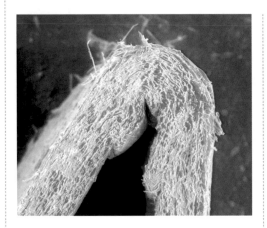

• Photo taken using an environmental scanning electron microscope (ESEM) at a 120x magnification, showing the damage done to the fibres in cardboard when folded. R. Passas – EPPG-PAGORA microscopy service.

Making Paper

In his report "Voyage au coeur du papier", Chrisian Voillot of the Ecole Française de Papeterie et des industries Graphiques tells us that "paper is a complex material made from mineral products and millions of fibres, each of which determines its properties. Plant fibres from wood comprise over 90% of paper. The remaining 10% come from other plants, such as straw, cotton, flax, hemp, esparto grass, kenaf, jute, ramie, abacá, bamboo and sugar cane (specifically its waste product, bagasse). It can also contain artificial or synthetic fibres, which are used for specific purposes such as for security features in bank notes; mineral fibres such as glass, which are put into paper that is used in filters; animal fibres such as wool and leather; and even remnants of tobacco leaves, which are used in the production of paper rolls. Wood fibres, either new or recycled, are sufficient for making sheets of the majority of paper in everyday use. However, for more specific uses, fibres from other plants come to the fore, with their different morphologies being used to develop specific properties. A sheet of A4 contains between 5 and 20 million fibres […]. The mineral products in paper have an important effect on the appearance of the paper's surface. Mineral fillers (kaolin, natural or precipitated calcium carbonate, titanium dioxide, talc and colloidal silica) are used to give weight or a particular coating to the paper, and are found especially in papers used for printing."

Artisanal Paper

Some artisanal papermakers use fairly unconventional production methods, making their paper out of a mixture of fibres and minerals. At the Moulin de Kéréon paper mill in Brittany, Jean-Yves Doyard, a papermaker, and Dominique Doyard, a bookbinder, are immersed in an exciting world of papermaking: "Flax and hemp are natural and durable fibres, and are the key ingredient in our artisanal paper. Papers made from these fibres are softly translucent, and are ideal for decorative use, for movable screens to divide up a room, and also in humid rooms such as shower rooms and conservatories. Each sheet is made by hand, and all of them are original and unique. Paper made from flax has a high level of mechanical strength due to the length of the flax fibres. Because the fibres are not particularly sensitive to ultraviolet light, this type of paper is extremely durable. Our flax-based Lin de Lumière paper contains a special resin that lets more light pass through it and makes it hard-wearing, easy to keep clean and able to withstand temperatures of up to 120°C. The paper has a flaxen beige colour that is complimented by decorative effects that are created by adding light-resistant pigments. It can also be decorated with calligraphy."

✏ **EFPG : École française de papeterie et des industries graphiques; http://cerig.efpg. inpg.fr/dossier/papier-materiau/page02.htm; www.efpg.inpg.fr/; sarl scop Papier relié, Moulin de Kéréon, www.moulin-de-kereon.net/; Le Moulin à papier de Brousses, 11390 Brousses-et-Villaret; Fibres libres, 56320 Lanvénégen; Le Moulin du Verger, 16400 Puymoyen**

• *Lamps made from artisanal flax paper with wicker reinforcements Designed by J.-Y. Doyard.*
Flower lamp, bedside lamp and large origami lamp.

Folded Books

BRIG LAUGIER

For over twenty years, Brig Laurier has had an astonishing and unusual artistic passion: folding books. Brig's precise and meticulous approach creates some remarkable pieces of work. "My work as an artist focuses on experimenting with the expressiveness created by folding book pages. My materials of choice to do this are books themselves; bound volumes whose spines create the folds that create a page-by-page rhythm. I work towards creating 'archi-texture.' I give real volume to a bound volume; I turn it into a sculpture. Folding book pages is different from folding a loose sheet of paper, because there are more restrictions on the folds you can do. The folds also have to be more repetitive, as it's the ensemble of the pages that creates the form of the piece. In folding a book page the text is being turned in on itself, making it into a secret. Yet, paradoxically, in the end the book takes on a completely new, open form in which it outwardly projects a new typography. It becomes dynamic, its freed pages swaying when you hold it. Folding creates movement. The form of the piece rises out of itself."
REMARKS FROM 22 JANUARY 2008.

• *Detail of* Page tournée.
• Libre prière 1.
• Libre prière 2.
• Contre vents et marées.
• *Detail of* Contre vents et marées.

Pop-Up and Movable Books

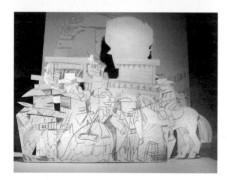

Show a child a pop-up book and he or she will be amazed by its moving pages. A magical world is opened up as the child pulls on the tabs that make the folded figures of the book come to life. Massimo Missiroli, a paper engineer and author of a number of pop-up books, and Patrizia Ghidarelli in the glossary of *Il libro ha tre dimensioni* have this to say about the production of pop-up books: "The content of pop-up books rises up from the pages when the book is opened, before folding back in on itself when the book is closed. The simplest pop-up effects are created simply by cutting around images printed on pieces of card and folding them. In more complex books further elements are added and fastened in position. For a pop-up book to work properly, its structures need to be directly connected to the two adjacent pages of the book, because it is by opening the book that the three-dimensional effect – and sometimes movement – is created. With the exception of carousel books, and according to the particular technique of each book style, the 3D effect is created by positioning the facing pages of the book at either a 90° or 180° angle."

The pop-up technique has been used since the fourteenth century in anatomy books, with pop-up images of the parts of the human body serving an educational purpose. But it was only in the nineteenth century that pop-up books for children started to be created. In the 1960s and 1970s there was an explosion in the popularity in the style, especially within the American publishing industry. One of the most famous pop-up books of this era, Jan Pienkowski's *Haunted House*, a striking piece of work that incorporated a wide range of pop-up techniques, was published in 1979 and went on to become a best-seller. Today, pop-up techniques are widely used in marketing and advertising, and are enjoyed by adults almost as much as by children

• *Carte de vœux by Jean-Paul Moscovino, 2008.*
• *Philippe Hugier, known as UG, **Morse**, édition, 2007.*
• *Philippe Hugier, **Big City**, model made for Gaumont for the film version of **Big City**, 2007.*
• *600 pastilles noires by David A. Carter, Gallimard.*

• *Marion Bataille, ABC 3D, letter "U",
a remarkable and creative alphabet primer,
éditions Albin Michel Jeunesse, 2008.*
• *Philippe Huger, Cristal, édition, 2008.*

• *Trogligami by J.-Ch. Trebbi, 1996.*

Trogligami

Trogligami involves applying pop-up techniques to simple cut-outs to create light and shadow effects and in so doing capture the essence of primitive dwellings. The contrasting surfaces of rocks create striking changes of shade that are difficult to recreate with just a piece of paper. But the main idea of the pieces is to simulate the spatial dimensions of these dwellings and in turn let the viewer's eyes wander and imagine.

Origamic Architecture

Masahiro Chatani, a Japanese architect and lecturer at the Tokyo Institute of Technology, is a pioneer of "origamic architecture", a style he created in 1981. His technique is recognizable for its purity of lines and the incredible finesse of its models. He has created numerous works of origami whose designs are often borrowed from the world of architecture. He has put on some magnificent exhibitions of his work and has influenced numerous other artists, such as Takaaki Kihari, who specialises in creating giant pop-up art, and Ingrid Siliakus, who has created some magnificent pieces that were also inspired by the works of the celebrated Dutch graphic artist Maurits Cornelis Escher (1898-1972).

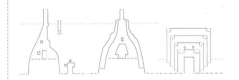

✏ **Robert Sabuda and Matthew Reinhart; Patrizia Ghirardelli and Massimo Missiroli; Guylain and Thierry Desnoues, Jacques Desse, http://livresanimes.com/index.html; www.japandesign.ne.jp/IAA/chatani/; Keiko Nakazawa; http://www.japandesign. ne.jp/HTM/TORYUMON/Db/kihara/eng2. htm; http://ingrid-siliakus.exto.nl/;**

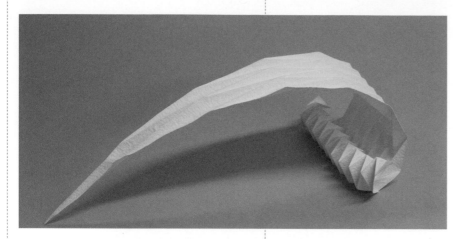

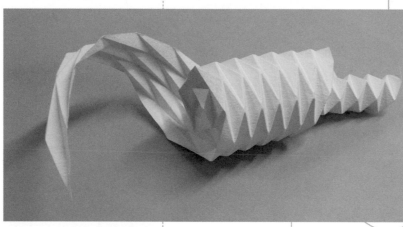

• *Paper sculptures by Luisa Canovi:* Ponte, Foglia, Fuoco.

Geometric Folds

Many scientists and mathematicians have taken an interest in geometric folds. Whether in the form of traditional Japanese tatogami (a type of small bag or pocket used for various purposes, also known as tatogami), tessellations or constellations, a world of shapes and forms appears before our eyes. It would be impossible to mention all the individuals involved in research into this field, but certainly one of the key figures is Shuzo Fujimoto, the man widely regarded as the father of modern geometric folding. Figures such as Toshikazu Kawasaki, Jun Maekawa, Thomas Hull, Robert J. Lang and Eric Gjerde have subsequently worked with tessellations, focusing on repetitive forms, symmetry and/or rotations, one example from this latter area being the twists of the works of Chris Palmer, who draws his inspiration from Arabic mosaics. Two further examples of the uses of geometric folds that are shown here are from Luisa Canovi, an Italian origamist, and a version of a dome piece based on a model by the architect J. S. Lebdev, the engineer V. F. Zdanov and the designer E. J. Bulgakova (In IL 32, 1983). Canovi, a talented designer, has produced a variety of creative works, including origami sculptures. She devises a network of folds that are based on geometric designs to create the desired curvature. But unlike traditional origami pieces that are not based on pre-set layouts and are built in successive steps, the contemporary technique is thoroughly pre-planned and then generally folded in one go. The two dimensions of the sheet are directly transformed into the three dimensions of the sculpture. According to Canovi, "although seemingly unrelated and based around different techniques, the traditional crane and my sculpture are in fact very similar, both being types of origami. There are admittedly dissimilarities between them, but on a different level. The differences between traditional and contemporary origami are the same as those between figurative and contemporary art. Just as the desire of the artist shifted from copying reality to searching for reality itself, so too has origami evolved from depicting a flower or an animal to evoking these things, without actually being them. It is said that traditional origami was passed from master to student. The complexities of some current models have made it necessary for the designs to be committed to paper to make sure they are accurately preserved. A new freedom of expression, decoupled from the principles of the rules, has led to the emergence of the contemporary origami *autcur.*"

Excerpts from De la tradition au design, Luisa Canovi, June 2005.

✉ **Eric M. Andersen: www.paperfolding.com/math/; http://mathworld.wolfram.com/Origami.html; http://kahuna.merrimack.edu/~thull/; www.paperfactory.it; www.JoanMichaels.Paque.com**

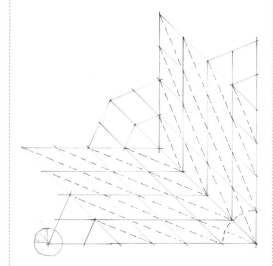

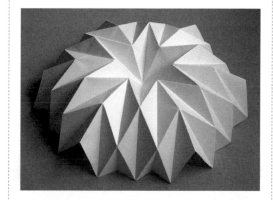

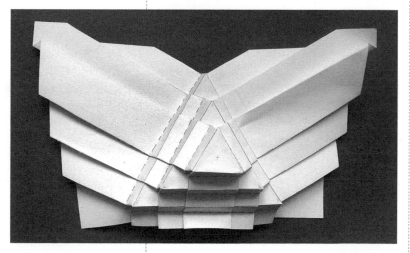

• Coupole, *folded by J.-Ch. Trebbi.*
• Pyramide : *designed and folded by J.-Ch. Trebbi.*

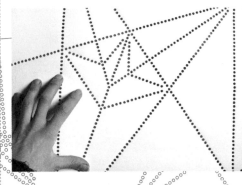

• Containers made from felt and paper. Punch-perforated. Leather fastening, 2006.

• Pattern for paper container. Punch-perforated paper, 2006.

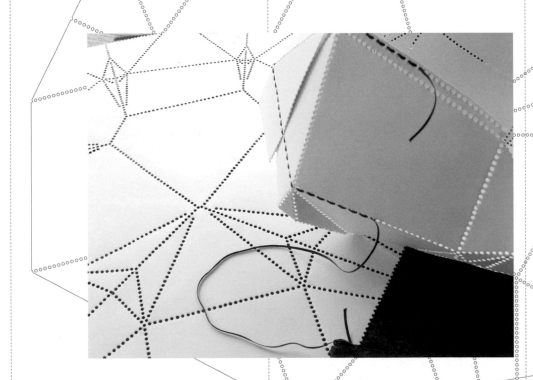

Paper Sculptures

MARIE COMPAGNON

Marie Compagnon's creative work involves creating 'perforated paper' shapes. "The process involved creating a two-dimensional pattern for a dodecahedron (a twelve-sided polyhedron) on a square sheet, and then taking the pattern and reshaping it into different forms using folds and counter-folds, whilst also keeping the 'offcuts' of the pattern, to create something random and unpredictable.

The idea was also to make the creations in two steps, with the patterns being automated and the assembly then happening by hand in a more spontaneous fashion. For my series of containers, first of all I modelled the shape on a computer. Then I unfolded it to get a two-dimensional pattern, all the while keeping the 'offcuts' between the faces of the shape.

The folds are marked out with a series of perforations, which act as a guide also allow the pieces to keep their shape and rigidity by passing a thread through them, rather like a seam. Here there was also a two-fold creation process: firstly cutting out the pattern, which can be done using automated processes, and then the manual assembly.

This production process can be used for other sheet materials such as felt."

🖱 **www.mariecompagnon.canalblog.com**

Punch-perforated paper, 2006

COLOURS, FOLDS AND SCULPTURES

HERE ARE THREE ARTISTS OF DIFFERENT NATIONALITIES: BRUNO MUNARI FROM ITALY, JEAN-PAUL MOSCOVINO FROM FRANCE AND EDWIN WHITE FROM THE USA. THEY ALL USE THE SAME WORKING TECHNIQUE, NAMELY USING PAPER AS A PLIABLE AND FLEXIBLE INTERMEDIARY MATERIAL BEFORE THEN CREATING A FINAL VERSION OUT OF METAL OR BRONZE. THE FIRST OF THESE ARTISTS CAME UP WITH TRAVEL SCULPTURES, A CONCEPT THAT ALLOWS PEOPLE TO CARRY DECORATIVE ART AROUND WITH THEM. CHANGING TO A BIGGER SCALE, THE SECOND CREATES OUTDOOR SCULPTURES. AND THE THIRD MAKES BOTH SCULPTURES AND MOBILES.

Bruno Munari:
• Travel sculpture made from cardboard, one-off piece, 17 x 47 cm, 1958.
• Model made according to the plan of the invitation card released on Munari's 100th birthday by the CLAC Galleria del Design e dell'Arredamento di Cantù (Como, Italy).

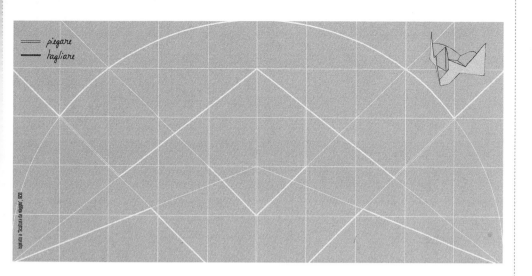

Travel Sculptures

BRUNO MUNARI

The Italian artist Bruno Munari (Milan, 1907-1998) was a prolific creator who straddled several different fields, including graphics, advertising and design. He was a painter, designer and illustrator who, in addition to producing around thirty children's books, also famously created "travel sculptures". In the February 1989 issue of *Domus* magazine, Marco Romanelli recounts the meaning attributed to these sculptures by Munari himself:" the sculptures are designed to bring a familiar cultural reference point into an anonymous hotel room or another place in which one might be staying."
His technique involved initially working with lightweight card, before then re-making the piece using square meshing, which allowed the preliminary model to be reshaped in a more intuitive way.
His sculptures were often made from cardboard, 3mm-thick aluminium sheets or wood; joints made from fabric or nickel-plated steel allowed the sculptures to take on different configurations. Munari liked to remark that "when objects that we use in our daily lives and the spaces around us become works of art, we will be able to say that we have achieved a new harmony in our lives."

✎ **www.munart.org; Galleria l'Elefante: www.galleriaelefante.com**

• *Creasepattern for an invitation card*
• *Travel sculpture made from weathering steel, one-off piece, 6 m x1,50 m x1,50 m; exhibited at Riva del Garda (Trento, Italy). Designed in 1958, created in 1997. Published with the kind permission of the Galleria Elefante, Treviso.*

- *Jean-Paul with a prototype in progress.*
- *Dialogue 2001, blue on aluminium, 180 cm height.*
- *Inter-Valle*
- *Tête dans les nuages, Tête en l'air, Tête ailleurs, Vol de nuit, 2005, blue on aluminium, 200 cm and 230 cm height.*

Folded Colour

JEAN-PAUL MOSCOVINO

Jean-Paul Moscovino deftly puts his capabilities as a sculptor to use in turning designs initially created on card into astonishing, folded contemporary sculptures made from steel and aluminium.
"The combination of my interest in architecture and classical design and my experimentation with different printing techniques have undoubtedly 'formatted' my artistic eye and the way I sculpt. My use of folding is the product of a long process of trial and error, and I have not really studied origami and its traditions.

When I was young, I used to help my father, who was an architect, as he produced models out of card-board. What interested me about them was the interplay of structures with their circulation spaces, and I liked to examine the way spaces were divided up.
Later on, I learnt design skills in a studio where an emphasis was placed on the built form, how to set out changes to plans and the layout of a design in accordance with the spatial limitations of the page. Using print methods in which the image that is printed is the opposite to the one on the print frame means that throughout the creative process you are continually working with an inverse image. Then I discovered screen printing and I plunged into the world of colour! I immediately fell in love with the thickness of the different inks and the vividness of their colours. I discovered that the colours that you print onto the paper showed through on both of its sides once the printing frame was lifted off.
I think that this was my starting point for my experimentation with the role of colour and the way it is perceived.

From the moment I began working with sculpture I started to make use of colour, covering all of the material I used in it. After using sheets of earth to create a sort of shell-like 'cover' of the hollow figures I made, I used thin wooden boards and then paper and card, which I found to be much more flexible and easier to shape. I create the forms of my pieces and develop a sort of blueprint that is then recreated on a much bigger scale out of a sheet of steel or aluminium, the choice of metal depending on the type of sculpture and whether it will be placed in an indoor setting or in a garden. But the material is really just a canvas and a vehicle for the colour.

Working with folds always presents a challenge. I always start out with a plan and a single sheet of material, and from the very first fold I transform it into a three-dimensional shape. Working in this way gives you a real feeling of there being two creative worlds at your fingertips and of creating two sculptures – a shape and an empty space – at the same time, with the space moulding the shape and vice versa. Because I have to take both these worlds into consideration at all times, and can switch between them at any time, I find myself in between the two worlds, variously in unstable equilibrium with them, touching them or on the edge of them. Through this dynamic, form and counter-form are created bit-by-bit. The later folds are more complex to achieve because the two different forms are as demanding as one another. Any move I make causes a chain reaction, with angles changing and clashing with one another; everything is in a continual state of flux.

I came to realise that my creations are defined by colours: I cut, fold, shape and crease them! Using this phenomenon as a material brings ambiguity, it reveals new things and it pushes the viewer to consider the interior and exterior forms of the pieces. And even if my sculptures are not 'habitable' in the sense of their form, they are all open and invite you to enter them. During the process of making them, I immerse and surround myself in colour!

• *Production phases of* Masque.
• *Sénégal 2003, yellow on steel, 80 cm height.*

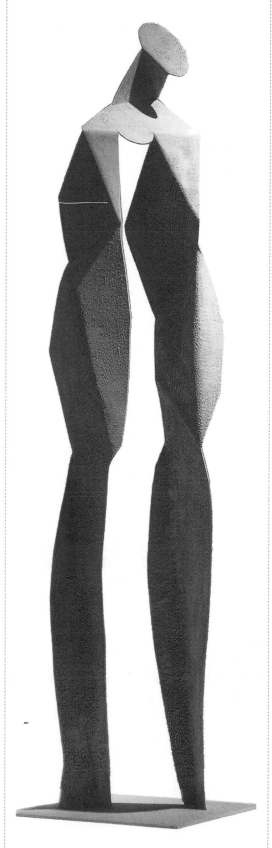

I've often used a matt cobalt blue in my works. The properties of this colour make it ideal for the spatial layout of my designs, and its uniformity does not make one's eye dwell on the texture of the pieces. Light and shadow endlessly play off against one another and deceive the eye. The spectator's perceptions of concave and convex change as you play with different colours. The reality of things is transformed by lighting: a strongly lit concave surface becomes convex, and shadows smooth over protrusions. Although the starting point of my pieces is often the body and anatomy – I anchor my pieces in reality – I also escape from these forms through folds, with their geometry combining with the forms to make them 'cubist'. This allows the pieces to take on an illusive quality as the forms that emerge are disconnected from the original idea.

Folds allow me to open up forms and to go inside them; they allow me to stimulate my creativity by forcing me to develop strategies for working with the situation they impose on me, a bit like playing a game of chess. I am also interested in the way the paper keeps a sequential record of the creases and also any mistakes that are made, meaning you can endlessly go back and revisit its process of creation." REMARKS MADE IN THIELLAY, 28 APRIL 2008.

🖙 http://membres.lycos.fr/jpmoscovino/

Sculptures and Mobiles

EDWIN WHITE

Edwin White, a member of the International Sculpture Center, lives in Siler City, North Carolina. He's practiced a range of crafts over the years, having worked as a freelance graphic designer, a designer of homes and studios, and as a carpenter. But in recent years he has mainly devoted his efforts to creating sculpture. He makes pieces cut from a very wide variety of materials that includes galvanised sheet metal, copper, stainless steel and black steel. The pieces are sanded and given multiple coats of exterior paint to make them as durable as possible. The originality of White's sculptures lies in their curvilinear style and his use of incisions, which gives them a moiré-effect transparency that is usually difficult to achieve with metal. White's approach has similarities with pop-up techniques and kirigami, which involves making cuts and incisions. Whether small or large in size, even the slightest air current will make the mobiles turn, creating a moiré effect through an interplay of light and lines.

"My fascination with origami, combined with my experiences of graphic and product design, has often encouraged me to work with metals as though they were paper. I usually use rolled metal and I make multiple cuts, often parallel ones, as well as perforating and manipulating it, to create the form of the object. The process creates forms that have an innate simplicity, something that is difficult to achieve in my other creations, which are usually made through welding or through a mixture of assembly techniques.

It was when I started to 'enlarge' my pieces that contain multiple cuts that the moiré effects came to the fore, something that I now try to incorporate into my work. The creative process begins with making models out of paper or polyester film. I find it much easier to create by cutting rather than using a paper sketch. The process of cutting and folding gives me insights that are very useful for planning the process of creating the form out of a larger piece of metal and ensuring that it will be stable. The bigger my creations are, the thicker the metal that I choose to make them from must be. As a result, more force must be applied when shaping the piece. Nowadays I use hydraulic cylinders and a selection of customized tools to do this. In terms of the tools I use, my most important innovation is a pneumatic tool mounted on a steel rod. This new device provides me with the necessary force to shape the 'ribs' of my pieces, as well as letting me work with steel plates in addition to sheet metal."
REMARKS FROM MAY 2008.

✏ **www.edwinwhitedesigns.com**

- *Left:* Sterling's grasp 2, *stainless steel, h: 2.74 m.*
- *Edwin with the model for his* Four Seasons, *stainless steel, h: 33.5 cm.*
- *Mengembang, Kuala-Lumpur Hilton, Malaysia, titanium, h: 5.18 m.*
- *Fem-Form, painted steel, h: 64 cm.*

GIANT METAL SCULPTURES

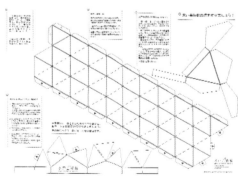

Art Tower Mito (ATM)

Located in the Ibaraki prefecture to the north of Tokyo, this contemporary arts centre, which comprises a theatre, concert hall and art gallery, was designed by the architecture company Arata Isozaki & Associates. The ATM is managed by the Mito Arts Foundation and hosts cultural and artistic events and activities. The complex is dominated by its very novel tower, which symbolises movement from the past to the future and from tradition to creation. Constructed on a metal framework, the tower rises to a height of 100 metres over the complex's plaza. It is made up of 57 1.5m-thick, triangular titanium panels. Each panel measures 9.6m across, and depending on the day's lighting conditions, the reflection of light off the panels produces a kaleidoscopic effect. This tower-cum-sculpture has four levels, including an observation deck.

✏ www.heatherwick.com; www.arttowermito.or.jp

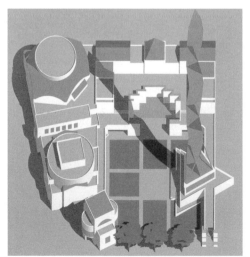

Vents

HEATHERWICK STUDIO

This project was conceived during the redevelopment of Paternoster Square, near to St. Paul's Cathedral in London. The square consists of buildings that have been constructed around an underground electrical substation, which needed a ventilation system. The client didn't want a single, large outlet on the grounds that it would turn the surrounding space into a corridor. After discovering two holes in the paving slabs over the substation, Heatherwick Studios proposed using them to create two outlet vents. This solution meant that the outlets would take up much less space. The 11-metre high mirror image sculptures were devised after experimenting with folded paper. The vents retain the same dimensions of the A4 paper that was used in the models. The structure is made from 63 identical, 8 mm-thick, isosceles triangle-shaped, stainless steel panels that are welded together and were given a satin finish through blasting them with glass beads.

MATERIALS AND TECHNIQUES FOR LIGHTING

ALTHOUGH LAMPSHADES ARE TRADITIONALLY MADE FROM PAPER, THEY CAN ALSO BE MADE FROM OTHER MATERIALS IN SHEET FORM, SUCH AS FIBREGLASS, POLYCARBONATE, PVC, POLYPROPYLENE, PARCHMENT OR RESIN-TREATED FABRICS. WHETHER THE INTENDED LOOK IS ANGULAR FOLDS, CURVES OR SLITS, ANYTHING IS POSSIBLE, WITH THE END RESULT DEPENDING ON THE CHOICE OF MATERIAL AND THE DESIRED LEVEL OF TRANSPARENCY AND OPAQUENESS. MANY DIFFERENT LIGHTING DESIGNERS HAVE USED A WIDE ARRAY OF FOLDS IN THEIR PIECES.

Le Klint

One of the most celebrated pieces of décor is unquestionably the modern and yet classic Le Klint lamp shade range. Made by hand at the company's headquarters in Odense, the history of the Klint begins in 1901, when the architect, engineer and craftsman P.V. Jensen-Klint created a pleated lampshade for a stoneware oil lamp. The Klint family soon turned to producing lamp shades by hand, with the company evolving over the years as new and talented designers, including Klint's sons Kaare and Tage and his grandson Jan, entered the fold. In 1950 the architect and designer Andreas Hansen added new models to the range, as did Paul Christiansen in the 1960s. Before long, the art of creating folded lamp shades had turned into an industry. The range of pleated and folded Le Klint pieces, made from plastic and paper in conjunction with a wide range of materials, including wood, iron, brass and steel, provide a nuanced selection of different shades.

• Classic Le Klint models: KL152, by Hvidt and Molgaard; KL157, by Andreas Hansen; KL 47, by Esben Klint.

Akari

The Japanese-American designer Isamu Noguchi (19084-1988) is known, amongst other reasons, for his "Akari" lamps, which are made from kozo bark paper and combine metal with bamboo. Dreamt up in the 1950s and 1960s and copied throughout the world, these flexible and lightweight sculptured lamps are inspired by traditional Japanese lanterns.

"Just like sunlight, the light of the Akari is filtered through shoji paper. The harshness of electric lighting is therefore transformed through the magic of the paper into the original light – sunlight – so that its warmth continues to fill our homes at night." (Isamu Noguchi)

Noguchi also experimented with designing pieces of furniture, such as his folded aluminium 'Prismatic Table', created in 1957 for ALCOA (Aluminium Company of America).

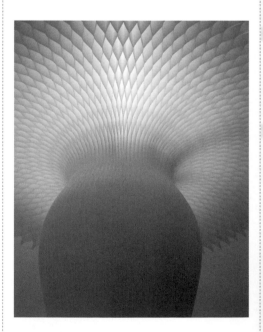

• Honeycomb.

Honeycomb

KOUICHI OKAMOTO

In 2007 the Japanese designer Kouichi Okamoto, of Kyouei Design, created the foldable Honeycomb lamp. This simple design can be folded and compressed to a thickness of 2 cm, and when unfolded is held in position with two small clips. It is made from denguri paper, a specialist product of the Shikoku region.

Pleated Lamp

INGA SEMPÉ

In 2001, Inga Sempé designed a stretchable lamp using a metal frame, a pleated textile lamp-shade and a halogen light source. Depending on the user's preferences, the height of the lamp can be adjusted between 45 cm and 2m, with the level of light provided varying in accordance with the length to which it is extended. In 2002 Sempé created a large pleated lamp for the Italian company Capellini.

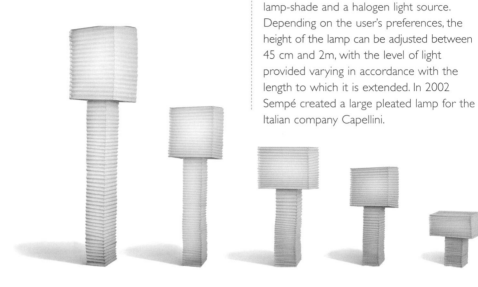

• Inga Sempé, *pleated textile lampshade, metal frame, halogen light source.*

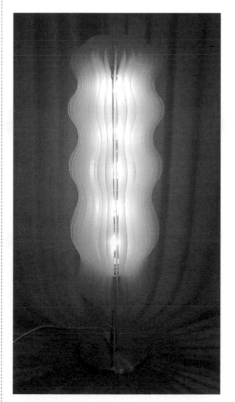

Honow Lamp

ISAMU TOKIZONO

Tokyo resident Isamu Tokizono makes his hanging Honow lamp from a single polycarbonate sheet into which parallel, equidistant slits are made. Due to its flexibility, the effect of gravity causes the lamp to take on original and inviting forms.

• Honow lamp - cedar pendant, *polycarbonate and cedar. By Isamu Tokizono .*
• Kagerou table. *By Isamu Tokizono*

✍ www.brave-design.com; Vanessa Battaglia, studiomold; Jean-Louis Bigou; Décoformes; www.ingasempe.fr/; Ramin Razani, Zoo lamps; www.origami.fr/; Andrea and Robert White; Daniel Lafer; Gonzalo Prades, lamps and recycling, radia'lamp; felt lamp by Mary-Ann Williams.

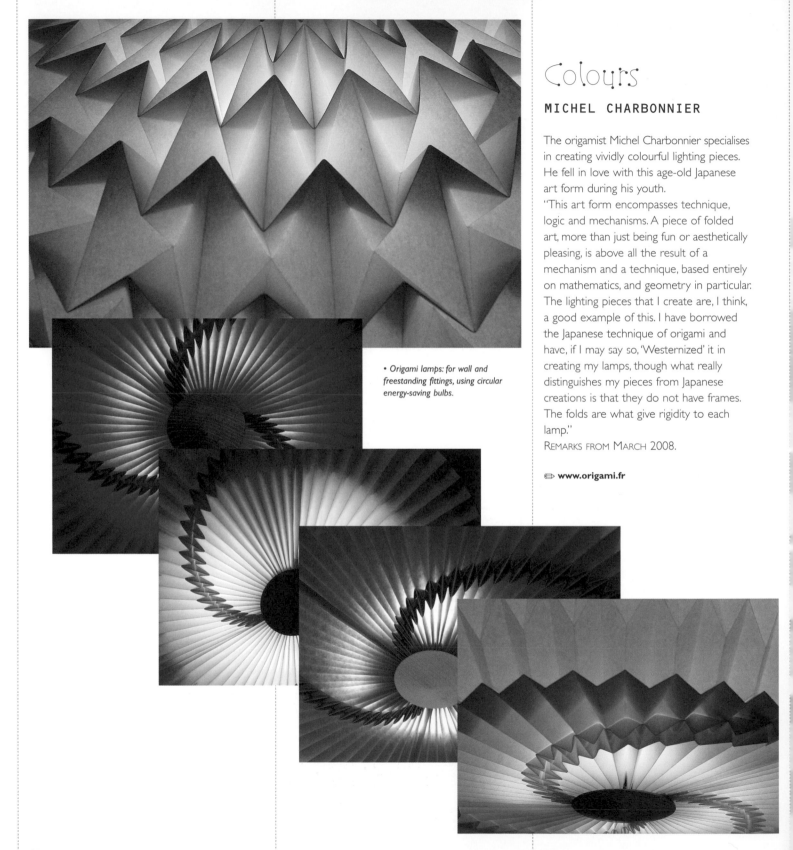

Colours

MICHEL CHARBONNIER

The origamist Michel Charbonnier specialises in creating vividly colourful lighting pieces. He fell in love with this age-old Japanese art form during his youth.

"This art form encompasses technique, logic and mechanisms. A piece of folded art, more than just being fun or aesthetically pleasing, is above all the result of a mechanism and a technique, based entirely on mathematics, and geometry in particular. The lighting pieces that I create are, I think, a good example of this. I have borrowed the Japanese technique of origami and have, if I may say so, 'Westernized' it in creating my lamps, though what really distinguishes my pieces from Japanese creations is that they do not have frames. The folds are what give rigidity to each lamp."

REMARKS FROM MARCH 2008.

✆ **www.origami.fr**

• *Origami lamps: for wall and freestanding fittings, using circular energy-saving bulbs.*

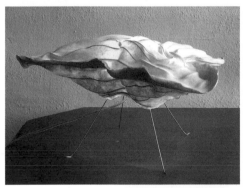

• Modular ceiling lights
• Dreispitz, Stèle, Huitre models, 80 cm
and Huitre ovale, 75 cm.

Lighting by La Font du Ciel

HELMUT FRERICK

Helmut Frerick is a paper maker and the founder of La Font du Ciel, a company that creates artisanal lighting pieces. In a perfect synthesis of ecology and design, Frerick makes his pieces from natural and plant materials, which help to produce a soft and relaxing light. "La Font du Ciel's creations are made from a hemp-based vegetable parchment. In designing and producing the lamps a strict environmental philosophy of making use of natural materials, producing the lamps with mechanical processes that don't involve the use of chemical products and integrating energy-saving light bulb technology into the lamps is followed.
The production processes that we use give each lamp its own unique undulations and create the almost pleated appearance of the lamps. Because they produce a soft and gentle light, lamps made from plant materials are ideal for spaces designed for relaxation. We produce one-off or limited edition creations that use plants such as hemp, flax, mulberry or even plants from the Ardèche mountains, as well as contemporary creations.
Our pieces, which are widely regarded as collector's items, have won awards for both for their artistic qualities and for the ecological approach of their production processes."
REMARKS FROM NOVEMBER 2007.

✎ **www.lafontduciel-luminaires.fr**

• Totem *lamp, polypropylene, energy-saving bulb, height: 70 cm. Designed by J.-Ch. Trebbi 2007.*
• *Lamp, height: 30 cm.*

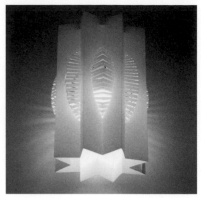

Orilum

JEAN-CHARLES TREBBI

"Luminous folds from the sharpened folder
The soft ambience of the evoking sculpture
Tactile pleasures of material in tension.
And culture structured by the folder's passion."

"It's about the search for forms, for sculptures that are beautiful whether lit up or not. It doesn't matter whether they assert a presence, or whether they are naturally lit by the rays of the sun or with artificial lighting, like in uplighters or desk lamps. They transform the ambience of a space, making it serene or dynamic through the colours they emit. The pieces here are made using folds and pop-up style cut outs, with the incisions creating enchanting luminous effects.

Through organizing geometric shapes from folds and planning out how the material – often cardboard, Bristol board or sheets of plastic or polypropylene – is handled, the forms are made, with the way the folds are marked out and the pieces take shape depending on the material used.

What material can be folded and then unfolded without it leaving a mark? Its very structure is modified through this action of shaping it, this expression of the folder, first in scoring, and then in making the fold. The creative pleasure of origami lies in taking something flat and then creating in three dimensions."

ORILUM, 2007.

Antonangeli Illuminazione

AKAMINE HIDETOSHI

Designed by Akamine Hidetoshi in 1988, the Kaj belongs in the category of timeless lighting pieces. With a frame crafted out of nickel-plated metal, the body of the lamp is made from a single sheet of white printed polycarbonate that has been folded by hand fifty-nine times.

This magnificent creation has echoes of the work of Azumi Hideaki of the mathematics department of the University of Tohuku's School of Science, an origami innovator whose work includes research into spiral shapes.

✆ **www.akaminehidetoshi.com/**

• *Akamine Hidetoshi,* Kaj *pendant light and* Kaj *lamp, 1994.*

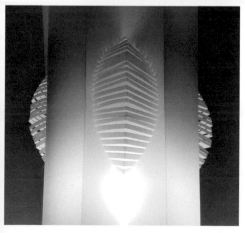

• *J.-Ch. Trebbi,* Petit Totem. *Prototype, height: 60cm, 2007.*

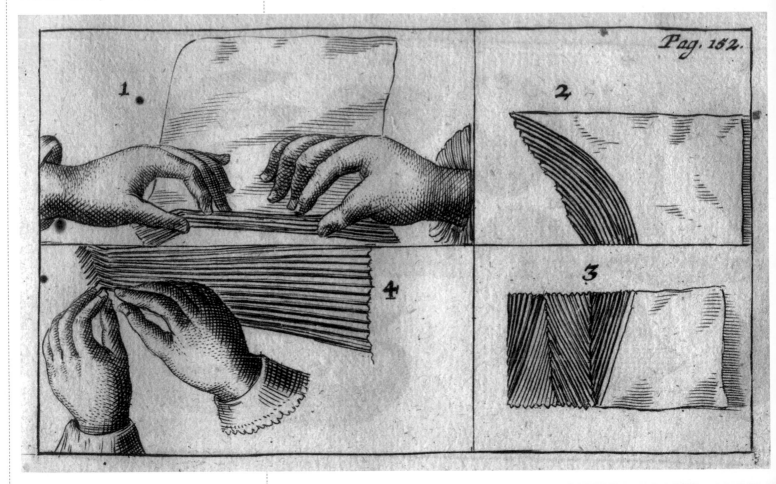

• Print of folding exercises for paper napkins, published by Andreas Klett in Wohl informirter Tafel-Decker und Trenchant *(Nuremberg, 1724)*. Source : Library/archive of the Paper Folding Documentation and Research Society (PADORE), Freiburg.
• *Italian lily, folded by Joan Sallas according to a design by Andreas Klett, 1677.*

The Art of Napkin Folding

JOAN SALLAS

Joan Sallas, a Catalan artist based in Freiburg, has spent many years carrying out research into origami and folding napkins. His work in this area has led him to argue that equating folding with origami — and in particular claiming that paper folding is a Japanese invention — is a somewhat lazy and inaccurate assumption. In fact, the Egyptians, Greeks and Romans also developed folding techniques, which can be observed in the clothing made by these civilizations. His research testifies to the diverse folding styles and techniques that were developed in the distant past in conjunction with paper folding. From the Renaissance onwards folding napkins has been learnt through first working with paper, before trying out the techniques on fabric.

Sallas has also studied numerous historical studies of the art of folding in Europe between the sixteenth century and the present day. These works reveal the high levels of imagination possessed by practitioners of the art during previous centuries. Sallas has presented some of the designs that he has encountered at exhibitions, including shows in Dresden in October 2007 and in Salzburg in June 2008.

• Peacock (Pfau) folded by Joan Sallas according to a design by Heinrich Louis Fritzsche, 1887.
• Fountain (Spring-Bronnen) made from folded napkins. Published by Andreas Klett : Neues Trenchier und Plicatur-Büchlein (Nuremberg 1677). Folded by Joan Sallas, for the Serviettenbrechen exhibition (2007) at the Kunstgewerbemuseum in Dresden.
• Ship (Nave) made from folded napkins. Published by Mattia Giegher (from Germany) in the Italian Li Tre Trattati (Padova, 1639). Folded by Joan Sallas, for the Serviettenbrechen exhibition (2007) at the Kunstgewerbemuseum in Dresden.

Soft Liquid(s)

DIANE STEVERLYNCK

"Soft Liquid(s) is a series of flexible containers that includes cups, bottles and decanters. Made out of a supple, waterproof cotton, these receptacles are unbreakable, lightweight and easy to store. The heavy-duty yet thin textile material from which they are made gives the containers both a soft feel and a high level of stability when they are placed on a surface. They are also malleable, changing in shape in accordance with the volume placed in them. The containers are made from a flat sheet and made entirely from folding, with no cuts being made to the material during assembly. Being assembled exclusively from folds means they are watertight and gives them an informal, simple and elegant look. Just like drinking water from cupped hands, the malleability and texture of the Soft Liquid(s) series lends a tactile and sensuous quality to the liquids poured into the containers. Shifting their shape under the pressure of your hand, the containers in many ways become a liquid material."

Production : prototype 2004,
material: watertight, folded
fabric, diameter: 8 cm,
height: 9.5 cm

• Soft Liquid

Orikaso Plates

Orikaso tableware can be carried anywhere, making it ideal for outdoor activities. These receptacles can hold liquids at temperatures of up to 100°C, and are also designed to withstand repeated re-folding at low temperatures. They are made from polypropylene, which, in addition to being regarded by Greenpeace as an alternative to PVC, is virtually the only plastic that can be folded without rupturing. Deezign-Mart offers further nifty kitchen and tableware accessories, including vegetable pickers and forks made from flexible stainless steel that fit on your fingers.

• Orikaso *after-dinner range*

NoMing Vase

FRANK KERDIL

When given flowers by a visiting guest, many of us are confronted with the problem of finding a vase for them. The NoMing vase makes this conundrum a thing of the past. Frank Kerdil is the brain behind this vase, which stacks flat and can be folded into shape in just a few seconds. The vase was developed in Denmark, where it is also manufactured. It is made from a unique, 100% water-tight material that is 51% chalk, which makes it much more environmentally friendly than regular plastics. It is flexible and unbreakable, not to mention re-usable and customisable. Dimensions: 10 × 10, h : 16 cm.

✎ **www.dianesteverlynck.be/; www.deezign-mart.com; www.noming.com/; www.sholk.com/**

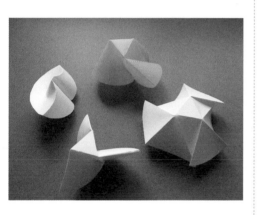

• *Orikaso.*
• *Orilum folded containers and vases by J.-Ch. Trebbi, 2007.*
• *Folding steps for the NoMing vase.*

FOLDED CERAMICS

THE IDEA OF FOLDING CERAMICS SOUNDS UNREAL. YET THE TWO TECHNOLOGIES DESCRIBED
BELOW ALLOW CERAMIC SHEETS TO BE WORKED WITH IN THIS VERY WAY.

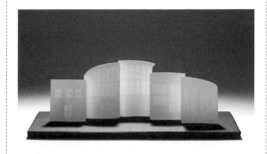

Arita Ceramic Paper

Inspired by a student who asked if earthenware could be used for origami, Hideya Seki, the President
of Kanyosha, a ceramics firm based in Aritacho, south-west Japan – a region well-known for its Aritayaki
porcelain – invented a sheet of white porcelain that can be folded like paper and can be used to make
origami.
After more than a year and numerous prototypes, the recipe for his invention, a mixture of ceramic
powder and Japanese washi paper, was perfected. The paper is soaked in a thick solution of ceramic
powder and is then left to dry until it turns into a sheet of white porcelain. Every sheet has a thickness
of between two and three millimetres and measures 15.5cm². After marking out the folds with a
pointed brush dipped in water and carefully making the folds, it is fired in an oven at 1250°C.

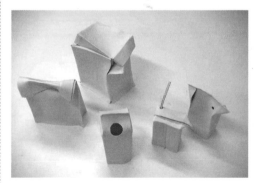

Kerafol Ceramic Paper by Keraflex

Kerafol is an advanced-technology porcelain sheet that was originally used in manufacturing the cooling
components of computers. The sheets are a light beige and have a thickness of between 0.5 and 1 mm,
and come in A4 (21 x 29.7cm) or A3 (29.7 x 42) format.
Oliver Vogt and Herman Weizenegger, two Berlin-based designers, experimented with these ceramic
sheets at the UPFOLD 1.0 atelier, creating all-in-one meal trays. The project was presented for the first
time in Portugal, at the 2003 Innova exhibition. Collaborative projects between the ceramic producer
Keramische Folien GmBH and art and design students from the University of Burg Giebichenstein in
Halle were also carried out between 2001 and 2002.
Some of the material's characteristics, such as its translucency and its fineness, open up the possibility of
bringing a contemporary and original feel to ceramics.

✏ **www.kerafol.com, distributed in France by Ceradel: www.ceradel.com**
andreamailen@freenet.de; anngbert@yahoo.de; Judith.Marks@burg-halle.de; Susanne.Bauer@burg-halle.de

*These pieces by students working under the guidance of Prof Hubert Kittel show the similarities
between working with ceramic sheets and working with card, such as cutting, punching,
stamping, folding, laminating and printing.*
• Andrea Nimtschke, **The City of the Worlds;** *materials: ceramic sheet,
0.4 mm and 0.8 mm thicknesses, laminated.*
• Annett Gebert, **Studies of Lamps;** *materials: ceramic sheet,
0.8 and 1.2 mm thicknesses, in strips.*
• Judith Marks, **Cup;** *materials: ceramic sheet; 1.2 mm thicknesses,
card, wood and silver.*
• Susanne Bauer, **Necklace;** *materials: ceramic sheet; 0.8 mm thickness, platinum and gold.*

Fans

SYLVAIN LE GUEN

Sylvain Le Guen is a talented French artist who has been interested in fans since his youngest years. He creates his fans one at a time, carrying out all stages of the production process himself, from choosing the materials and creating the motifs to folding the leaves and putting the finishing touches to the final product. All of his pieces are unique, and each one is signed and dated. Influenced by Japanese design, he draws his inspiration from origami, and has breathed new life into contemporary fans by trying to turn them into true pieces of artwork. His first exhibition, held at the Musée de la Mode, Paris, in 2001, showcased themes inspired by the fold-work of fans.

His conceptual approach is totally original, and is based on giving a new interpretation to the tradition rules of fan-making.

The methods he has developed have given birth to a galaxy of romantic styles:
- Ecornés: folds and counter folds made in the leaf are 'dog-eared' and the leaf has a creased effect
- Décalés: the leaf of the fan, made from just one piece of material, has incisions in certain places and rests on a monture with a twisted shape, giving the fan an 'off-kilter' look.

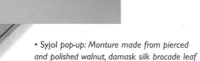

• Syjol *pop-up*: Monture made from pierced and polished walnut, damask silk brocade leaf with pop-up 'blooming' brocart flowers.

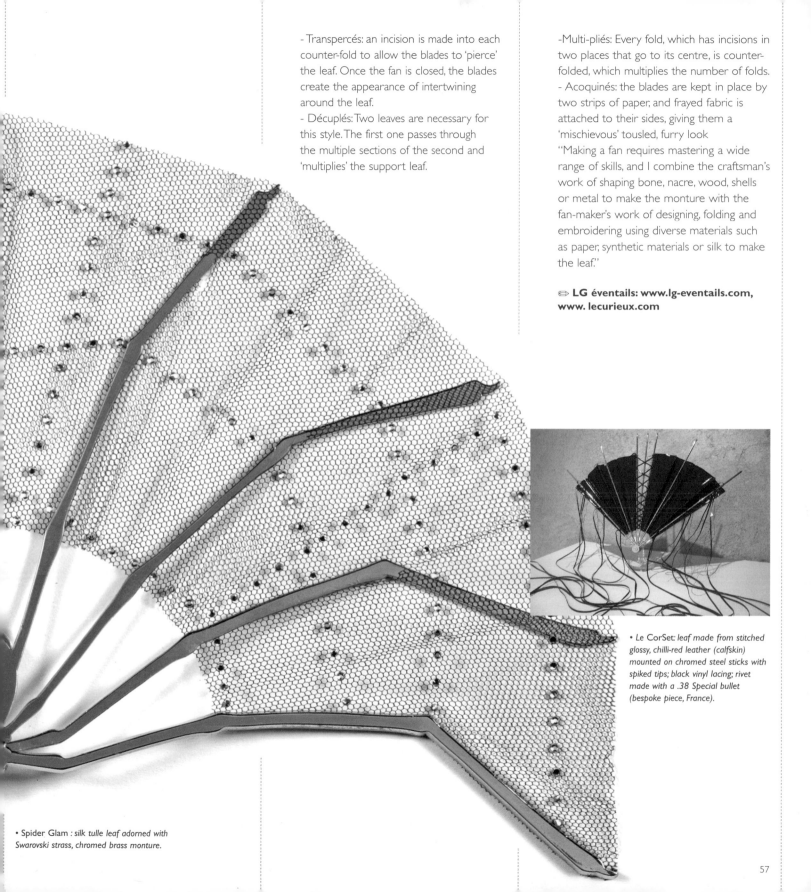

- Transpercés: an incision is made into each counter-fold to allow the blades to 'pierce' the leaf. Once the fan is closed, the blades create the appearance of intertwining around the leaf.
- Décuplés: Two leaves are necessary for this style. The first one passes through the multiple sections of the second and 'multiplies' the support leaf.

-Multi-pliés: Every fold, which has incisions in two places that go to its centre, is counter-folded, which multiplies the number of folds.
- Acoquinés: the blades are kept in place by two strips of paper, and frayed fabric is attached to their sides, giving them a 'mischievous' tousled, furry look

"Making a fan requires mastering a wide range of skills, and I combine the craftsman's work of shaping bone, nacre, wood, shells or metal to make the monture with the fan-maker's work of designing, folding and embroidering using diverse materials such as paper, synthetic materials or silk to make the leaf."

✆ **LG éventails: www.lg-eventails.com, www. lecurieux.com**

• Le CorSet: leaf made from stitched glossy, chilli-red leather (calfskin) mounted on chromed steel sticks with spiked tips; black vinyl lacing; rivet made with a .38 Special bullet (bespoke piece, France).

• Spider Glam : silk tulle leaf adorned with Swarovski strass, chromed brass monture.

• Lotus Noir, *folding fan: the leaf is made from black organza and pongee, silver and Swarovski strass sequins, chromed brass monture (Fan Museum collection, London).*

• Sultane Noire Diam's.
Monture: blackened, pierced and polished wood with a fine silver glitter finish. Strictly speaking this fan doesn't have a 'leaf'. The fan is constructed through stitching together the folds between the blades, which creates a decorative effect out of the perforations in the monture and the zigzag stitching that can be seen even when the fan is closed.

• Facing page: Dame de Lotus, *polished bone monture, ivory silk organza leaf, overlaid with three rows of stitched silk petals, giving volume to the folds of the leaf (private collection, Netherlands).*

• Bag, 2001, Polyester and leather fabric, plated, pleated and seeded.

Necklaces and Bags

TINE DE RUYSSER

A doctoral student at the London Royal College of Arts, Tine is part of a study group that is developing a material that that combines the look of metal with the suppleness of textiles, with the goal of finding applications for it in jewellery, architecture and other areas of design. She creates geometric though nevertheless poetic pieces of jewellery and other objects that are the product of complex interactions between fold patterns and the texture of the material, which is made from a mixture of polyester, leather and plastic.

✎ **http://petitsplis.cabanova.fr; l'Inédite Petits Fragments, origami jewellery ; www.tinederuysser.com/ Site/Home.html**

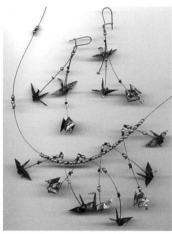

• Origami necklace and earrings by V. Wardega.

• Necklace 2004, Polyester, leather and plastic fabric, plated, pleated and seeded.

Origami Jewellery

VÉRONIQUE WARDEGA

"I was taught the art of folding paper at primary school. Then, over time, the little jumping frogs of my childhood were joined by the wild cranes, butterflies, lotuses, roses and lily flowers that I discovered in origami books. Having had a preference for working with smaller sizes (from 5cm squares), I turned my attentions to jewellery, encouraged by several decisive encounters that I had with other artists. My business, Petits Plis, was founded on the back of this in September 2006. The creases are varnished, which makes them permanent without diminishing their delicateness, the transparency of the paper, or above all their lightness. Depending on the beads that are combined with the origami, the necklaces, bracelets, brooches, earrings or hair ornaments can be variously sophisticated, refined or playful. I've also discovered an enchanting new form of recycling, in which metro tickets, yoghurt pot lids, sweet wrappers or teabags are transformed, as if by magic, into elephants or flowers! I have a strong connection with the spiritual side of origami, and so in September 2007 I created "oPAIXrcule",

a crown of a thousand cranes made from yoghurt pot lids, in homage to the garlands of cranes laid in Hiroshima to promote peace."
REMARKS FROM JANUARY 2008.

• Right-hand page: oPAIXrcule, by Petits Plis.

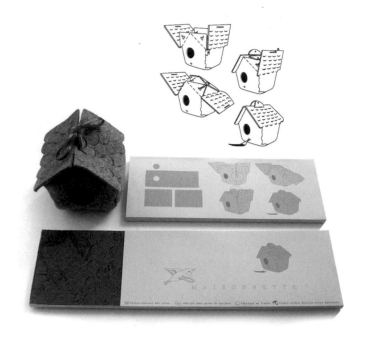

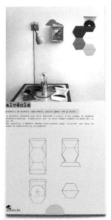

Creations by Studio Lo: bags, nest boxes, slippers, CD pockets

Accessories, Slippers, Bags

STUDIO LO, JOSH JAKUS

At Studio Lo, designers Eva Guillet and Aruna Ratnayake have developed their own special production methods. "With an awareness of the economic s realities of the design industry, we set up our own independent company in 2005. Our desire to produce goods using responsible methods led us to use natural materials and work wherever possible with a local network of SMEs. We developed a 'pauper's design' style that only requires a modest investment and is based around small-scale production runs. The research we carried out into materials optimization, customization and production chains led us to deploy water-jet cutting systems. As we make our pieces we endeavour to create expressive and clear production processes.

Slippers: made from a single piece of screen-printed felt, water-jet cut, assembled with fasteners.

The slippers are made using a contiguous design pattern, meaning two slippers are made using each cut. This principle of 'shared cuts' is one that we try to incorporate into our creations. Rotating the printing screen and using a 'shaky' cutting technique also allows us to make one-off runs of our designs. Printable honeycomb CD pocket, 140g grooved paper, prototype. When flattened out, the pocket is 29.7cm long and can be printed on and customised using a selection of templates or with the help of the software on the accompanying CD. A simple series of folds combine to create a pocket with an efficient closing motion. A double-sided tape fastening allows the pockets to create a modular, ultra-flat system of wall storage for CDs.

Natural grey felt nesting box, die-cut, assembled with fasteners.

We also apply our design principles to making objects such as this one. The roof of the box opens up to allow the interior to be cleaned. An ersatz perch can be fashioned out of objects such as forks, pencils or twigs.

Bag: Made from screen-printed felt, water-jet cut, assembled using zips.

The Opla bag stands at about a third of the width of a standard roll of felt. Its zip is like a detachable seam that is used to assemble the bag, and also opens up the possibility of adding 'peripherals' to it."

There are other folded fashion accessories that are worthy of mention, not least the range of coloured, Tetra carton-shaped felt bags produced by the Japanese label Tokyo Power, or Aiko Machida's refined bags made from folded leather.

The bags made by the Californian designer Josh Jakus, on the other hand, are made using a different approach. Jakus set himself two challenges. The first was to capitalize on the density, texture, strength and flexibility of felt to make objects from just one sheet of the material, and the second was to transform a flat surface into a three-dimensional object and vice-versa simply by using a zip. In 2005 Jakus created his UM bags using two layers of material made from 85% wool and 15% other fibres. In a similar vein, he creates objects in a way that takes advantage of the simplicity of their forms and the materials they are made from. For example, rather than simply being thrown away, the circle cut-outs that are created in making handles are used to make drinks coasters.

Experimentation with recycling PVC has led to the development of flip-flops made from Pévéchouc, a novel recycled material developed by Bernard Chaize. Pévéchouc is made from recycling leftovers from industrial processes such as the manufacturing of car dashboards and door trims. After processing in granular form, the material can be worked with using a variety of techniques, such as injecting, extrusion and calendaring. The flip-flops are shaped out of a simple fold in the material. The wearer's toes fit comfortably underneath the strap formed out of the fold. Portable and customisable, the blue flip-flops are made out of the cut-offs from swimming pool covers, while the black versions are made from leftovers from leather-making, car trims, lorry canvas etc.

✏ **http://studio.lo.neuf.fr/; www.trucs-trouvailles.com; powarch.com; www.aikomachida.com/; joshjakus.com/**

- UM bags *by Josh Jakus*
- *pévéchouc flip-flops*

Crumpled Clothing

ANNETTE AND PAUL HASSENFORDER

Drawing their creative inspiration from art, music and literature, Annette and Paul Hassenforder are creators and storytellers who are infused with a wondrous energy for passionately communicating their expertise.

Eschewing traditional dressmaking methods, Annette instead creates her garments through crumpling and the tensile properties of folds.

There are three crumpling patterns: random, parallel and radial. It's difficult to communicate these types of folding using diagrams, as being able to use them requires personal experience and savoir-faire.

According to Annette Hassenforder, "a crumpled piece of paper usually thrown straight in the bin. But if instead of doing that you hold it in your hand and look at it closely, you'll notice that it's covered in veins and arteries of all different sizes. And what seemed like a dead object suddenly comes to life, metamorphosing into animal or plant tissue. Likewise, if you take the trouble to examine a rose-bud, you'll see that the petals inside are crumpled and puckered, waiting to unfold and bloom. Having appreciated these textures the next step is to get crumpling, storing away your paper to unfold later on. In the hands of an expert folder, crumpled paper becomes a material of its own; it takes on a texture and invites you to create with it.

Paul Jackson, one of the great origami artists, demonstrated crumpling techniques at the MFPP's convention in Sèvres, and we also crumpled with the CRIMP team. For the last four years, we've been working on clothes made from crumpled paper, including dresses, hats, scarves and boas. In the process, we became "Créations Paula".

• *Crumpled texture.*
• *Bustiers and dresses.*

But why crumpled clothes?

Why? Why not?

Because they're pretty,

Because any woman who sees them wants to wear them,

Because to us paper is just as refined as the most beautiful fabrics,

Because it throws up so many challenges,

Because you can't cheat; if the forms aren't 'pure', the garment falls apart.

We refuse to treat paper like fabric, by which we mean that we refuse to use the dressmaker's technique of cutting out from a pattern and stitching.

We want to make clothes just through folding, so that the folds and their tension are what make the garments stay on the wearer."

🖂 **www.flickr.com/photos/7314460@N04/sets/72157603819122552/**

• *Hats by* **Créations Paula**.
• *Vagueline hat, random folds.*

Hats

JEAN-LOUIS PINABEL

Jean –Louis Pinabel invites us to share his enjoyment for the intricacies of pleats, which are used to delicate and thoughtful effect in his works :

"As a milliner I have had the opportunity to explore such varied worlds as felt, exotic types of straw and leather, flowers and feathers. But for me the world of pleating is an infinitively creative one. It has also provided me with new friendships: without the warm and invaluable support of the Maison Lognon team I would have only discovered a fraction of the pleat's unlimited creative possibilities.

The milliner makes use of bias and drape in his works. Pleats give life, texture and volume to even the most lifeless of materials, such as straw, hair, felt, leather and veils. And let's not forget the materials still to be tried out, in the never-ending quest to adapt the materials to the designs, and not the other way around, and to create surprising – and perhaps even impossible – designs!

Certain materials never create the same pleat twice. One would think that pleats would follow the mathematically precise style of 'tailored cuts' and would therefore be reproduced infinitely, but they also have a certain 'looseness' to them. Arranging the pleats in a different way can create a radically different rendering. You can't recreate Madame Grès's drapes. Trying to decode the magic of a vivacious pleat is useless: it's a divine mystery created by the algorithm that's encoded in the card, which infuses the material with spirit and sensuality, so the mould brings out a look in the silk that is like no other.

Through pleating, the most stubborn of materials becomes obedient, limpness is turned into energy, and neutrality into vibrancy. All this leads to a creation that has a supple bias, a graceful drape and tailored precision. Now we must honour women and crown them with works that combine precision, grace and suppleness."
REMARKS FROM MAY 2008.

• *Hair pillbox with accordion pleats. Jade and feather brooch with jet-black, small-chevron netting.*
• *Sunburst-pleated demi-pointe with organza drape with Italian straw brim.*
• *Flat-pleated straw turban.*
• *Flat-pleated sisal turban, accordion-pleated, straw stole.*
• *Right-hand page: Feather hat with frilled wild silk halo and grand chevron flat-pleated head-scarf.*

Pleated Textiles

ATELIER GÉRARD LOGNON — PARIS

Flat, sunburst, accordion, organ, crinkled, honeycomb, crumpled, fantasy, Fortuny, chevron and Watteau are just a selection of the highly evocative labels given to different types of pleating. But in reality, pleating does not have a standardized nomenclature. The forms of the pleats tend to speak for themselves. And in Gérard Lognon's atelier, pleating takes on a magical form!

"Atelier Lognon is located in the former Biou pleating atelier, which until the retirement of its last designer was a hub of haute couture run by a family with a long heritage of pleating. Going back even further, I believe that prior to 1914 it was a milliner's atelier. Until the 1950s, the Bourse and Quatre-Septembre neighbourhood of Paris was teeming with ateliers practicing a diverse range of fashion crafts, employing thousands of workers and artisans. There was Rue de Richelieu, where my grandmother worked in a milliner's atelier, and Rue des Petits-Champs, home of Maison Michel and Maison Legeron, known respectively for their hats and their flowers and feathers. It's almost all gone now.
Jean-Louis Pinabel, December 2007.

Textiles are a family affair for Gérard Lognon. His great grandmother was a linen maid in the household of Empress Eugénie, and Gérard has memories of the business stretching back into his childhood.
"I love walking into my pleating workshop because of the distinctive smell of cardboard that drifts through the place all year round. When I was a schoolboy I used to climb the stairs up to my parents' home, and I always knew I wasn't in the wrong place because at the time I arrived they had just turned off the steamers, and I could smell the steam and the wet cardboard. My interest in the craft stems from being able to work with designers who get us involved with developing their creations. We meet with haute couture and prêt-à-porter designers four times a year to examine new styles and prepare their designs for the coming season."

The majority of the 3000 moulds used by the Atelier Gérard Lognon were made by his grandfather. These tools are treated with the utmost care, and according to Liliane Leboul, the head of the atelier who has been lovingly working with pleats for thirty-seven years, are only used once a day. The pleating process is carried out by a two-person team. The pleating mould is made from two sheets of Kraft paperboard that fit together perfectly. It is positioned on the work table and opened to separate the two sheets.

The fabric (in this case chiffon) is then spread out onto the mould. The top sheet of the mould is then very carefully placed on top, in line with the markers. To prevent the fabric from moving, wooden slats that are approximately 10cm wide are placed on the top of the mould, held in place by weights.

Next comes the most delicate part of the process, in which the two-man team, standing at the end of the table and starting at opposite ends, set about shaping the pleats by hand, moving towards one another as they do so. The pleats are made at a perpendicular angle to the length of the fabric.

The moulds have a length of between 1.50m and 3.50m, and a maximum height of 1.40m.

The ensemble is then wrapped around a metal cylinder, before being given a protective layer of Kraft paperboard and placed in the steamer, a sort of giant, vertical pressure cooker that is heated by gas and can hold up to 10 rolls.

When pleating tulle and chiffon the steamer is set at 85°C and the materials are placed in it for 20 minutes, while woollens go in for 50 minutes at 95°C and polyester for 50 minutes at 105°C.

The protective paperboard is then removed from the mould. The fabric, still on the cylinder, is then cooled for at least 24 hours.

Sunburst pleats in chiffon do not undergo a rolling process, but instead are kept flat using slats
BASED ON A VISIT TO ATELIER GÉRARD LOGNON, DECEMBER 2007.

• A selection of Atelier Gérard Lognon's
magnificent moulds.

• Unlike accordion pleats, sunburst pleats are not rolled.
• The flat mould of a sunburst pleat.

• Removing a sunburst pleat from its mould.

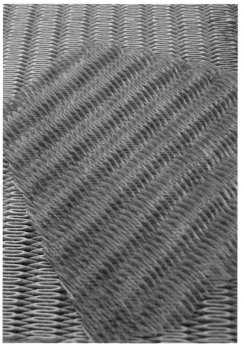

• Demonstrating the complexity of the pleating process by unfolding a mould.
• A mould and the pleat created from it.
• Steps for using the moulds:
– Holding the ensemble in place with weights;
– Covering the chiffon with a second sheet of Kraft paperboard;
– Rolling the ensemble around the cylinder before placing in the steamer.

Blinds

PIETRO SEMINELLI

"Folds determine the form and make it appear; they turn it into a form of expression."
Recently awarded the honour of *maître d'art* by France's Ministry for Culture and Communication for his innovative application of the art of folding to textiles, in each of his creations Pietro Seminelli draws on aesthetic and philosophical references borrowed from the Baroque. The infinite interplay of folds and counter-folds in the fabric as they interact with the light provides him with the opportunity to play on the confrontation between indoors and outdoors. Indeed, such is the subtlety of their design, his screens, which many regard as being 'stained glass textiles', seem to get their colour and be lit up 'entirely from within'. The Baroque influence is re-interpreted through compositions that have a restrained intensity and an almost contemplative moderation that verges on the symbolic.

The breadth of the expressive registers of his works, especially his textile creations, extends from designing costumes for an exhibition on the Miao people at the Musée des Arts Asiatiques in Nice to creating sculptured costumes for the Musée des Tissus et des Arts Décoratifs in Lyon (where Seminelli's works are currently on show), and even to sculpting monumental textile pieces. His creations can also be found in Chanel's boutiques in New York, Dubai, Ginza and Paris. Seminelli's collaboration with some of the major names in architecture and decoration, such as Peter Marino, Michael Graves, Frank de Biasi and Jamal Lamiri Alaoui have provided opportunities to develop and refine unique and functional designs. Seminelli's expressive folding is a distinctive feature of his creations— it is both his leitmotiv and modus operandi – and can be seen, for example, in the infinitely delicate translucent effects in his pieces. Conscious of the expressive possibilities of textures, Seminelli develops and produces his own fabrics from natural abaca, banana tree, flax, palm or ramie fibres, which are intertwined with stainless steel, brass, lurex or nylon threading, with the goal of creating fabrics with new and original translucent properties. Whether used in the fixed or sliding screens, blinds, curtains, shades or lamps that contribute to creating a timeless, characterful and expressive ambience, folding is not really a question of style, but rather a way of thinking.

"Ideas are folded within our souls, the depths of which are like a black drape, ridged with myriad pleats which are crossed by our perceptions; only a small fraction of them reach the threshold of our conscience."
In Le Pli – Leibniz et le baroque, Gilles Deleuze, Seuil, 1990.

✐ **www.seminelli.com**

• *Working on the model for folds in a linen and silk piece.*
• *Mould for a chevron pleat, made from strong Kraft paperboard.*
• *Section of a screen made from polyester fibre and metal, with a scale-effect folding pattern.*
• *Detail of a test sample containing variations of two hollow folds.*

Right-hand page:
• *Octagonal apertured pattern, tracing paper prototype, 1990.*
• *Central panel of a blind for a bow window, square twisted fold, lurex silk, 2005.*

Carapace
PIETRO SEMINELLI

"This piece is an expression of something that is very important to me. It's a personal exploration that I have been nurturing for the last fifteen years. Folding gives it a sense of mystery and magic. The notions I attach to folding are envelopment, skin, mutations or metamorphoses, chrysalises, secrets and intimacy. I'm unable to separate these notions from their bodily implications. In fact, whenever I put folds into cloth, fragments of my emotions and past escape into them. For me, life is a succession of metamorphoses, and each stage brings us closer to the essence of what makes us human, in the sense of what it means to exist and be what we are. Happiness and grief both manifest within ourselves and leave their mark on our whole being. Part of this labour of becoming permanent is transmitted in my piece "Vestito d'ombre". It's the dress of a warrior grappling with his demons, fears and battles: a fabric robe with folded scales made from bronze linen, and a carapace made from Miao fabric laminated onto paper."
REMARKS FROM FEBRUARY 2008.

• Vestito d'ombre, *fabric robe with folded scales made from metallized bronze linen, carapace back made from pleated Miao fabric laminated onto ink-washed paper, tarlatan coating, 2006. Background, pleated scale chevron drape made from Miao fabric*
• *Detail of a scale pleating in baptiste de lin fabric, creating rhythm and translucence.*
• *Sample of an elongated scale folding pattern.*
• *Close-up of panel, frieze of twisted squares, natural abaca.*

Curtains

HANNAH ALLIJN

Irritated by massive and heavy curtains that seemed to absorb light even when open, Dutch designer Hannah Allijn wanted to come up with a way to get rid of this bulkiness. The challenge for Allijn was to turn a large surface area into a more manageable and smaller one, as well as being able to play with positioning the curtain. Pulling the cord of the curtain makes the triangles fold together one-by-one, and as they do the curtain changes shape and comes to life. Geometric forms are created as the curtain folds up until it reaches the corner of the window. When the cord is released, a weight causes the curtain to unfold again.

✎ **www.allijn.nl/**

Furniture and Architecture

• *Hiroki Takada, Nike, 2007.*

• Bentwood furniture *by G .Candilis, an astonishing account of the Viennese bentwood furniture of Michael Tonet.*

BENTWOOD

The technique of bending wood using steam is an old and well-known one.
Michael Thonet, who began making bentwood furniture in 1841, is generally credited with developing this technique, in which steamed beech wood slats are shaped in accordance with the shapes required for the furniture. His technique marked the first attempts to produce curved wood using industrial methods. Thonet's pieces were often imitated and their shapes warped. Then, in 1932, the celebrated Finnish architect and designer Alvar Aalto, along with his wife Aino Mariso (also an architect) and furniture designer Otto Korhonen, developed a ground-breaking method for bending wood, which involved fitting strips of laminated wood into longitudinal slots. Gerald Summers's Plywood Armchair was then produced in London in 1934, from a single piece of cut and shaped plywood. Its simple, moulded design, a triumph of fluid contours, was achieved without the use of screws, nails or other metal fasteners. In 1945, Charles and Ray Eames had a vision for creating chairs for children out of moulded maple or ash plywood; the chairs are today produced by Vitra. In 1953, Charlotte Perriand unveiled her famous stackable chair made from moulded and polished wood. Nineteen sixty-three saw Grete Jalk, a Danish furniture designer, develop her elegant moulded plywood chair. Also worthy of mention are the moulded plywood chairs made from 1997 onwards by John Christakos's Blu Dot, which take the shapes of unusual insects and flowers.

Rather than being folded, wood is shaped and bent. To fold wood, it's necessary to use hinges and pivots. Different types of plywood such as multiply sheets or MDF can be bent to varying degrees.
For example, Plysorol's Flexply plywood panels are made from three or five tropical wood veneers, and come in 5, 7, 9 or 16mm thicknesses. A 5mm-thick piece has a minimum bending radius of 12cm. Some creations are enhanced by using specific types of wood, such as Capelle's Curévol composite wood, which is comprised of a polypropylene or cardboard honeycombed centre and exterior layers of tropical wood or other wood species, or MDF. In the past, wood was moulded in layered strips, which often became damaged during the crafting process. The defibrillated woods used today offer an incredible level of pliability.

BENDYWOOD

The Italian company Candidus Prugger is constantly working to improve its Bendywood. "Bendywood is still produced using a process that was patented in 1917. It is produced from hardwoods (such as beech, ash, oak and maple), which are cut into pieces with a 10 x 12 cm thickness and a maximum length of 280cm, and then steamed. The pieces are then compressed lengthways to 80% of their original length. The wood is then dried, still in its compressed form, with the 280cm piece now having a length of 220cm and a moisture content of 14%. The wood can be worked using traditional methods into hand rails, wood profiles for tables, slats for windows and skirting boards. The wood's special characteristics mean it can be cold bent up to a radius of 1:10. So, for example, a 22cm-thick piece can be bent to a radius of 220 mm."
Bendywood is used principally in areas of woodworking such as household furniture and other interior fixtures.
Thin pieces of the wood can easily be shaped by hand, but thicker ones need to be bent using machines such as roller benders like the ones used in artistic metal-working. Bendywood is used in a diverse range of situations, such as the skirting of round tables, mouldings and hand rails for staircases, when it would be too costly to use traditional methods.

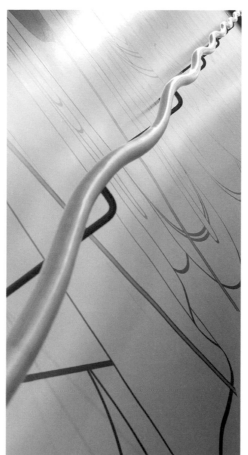

• Bending machine and applications of Bendywood in a bench and hand rail.

Carlo Mollino Furniture

CARLO MOLLINO

Carlo Mollino (Turin, 1905-1973), was an eclectic designer who practised as an architect, photographer and car and aeroplane designer, and was also a writer. He is known above all for his pieces of furniture, many of which were made using moulded plywood.

This highly imaginative and unconventional jack-of-all-trades genius was undoubtedly influenced by Antoni Gaudi, the Art Nouveau movement and Alvar Aalto. The majority of his large collection of furniture inspired by the forms of the human body is made up of one-off commissions.

His architectural works include the Regio theatre in Turin, the ceiling of which is made from pleated structures. The photos on this page come from a presentation of models produced by Fulvio and Napoleone Ferrari, based on sketches of furniture by Mollino. His approach was a simple one of cutting out from a flat piece of wood and using traditional wood-shaping methods.

✆ **www.candidus-prugger.com; www.bendywood.com; www.thonet.de; Plysorol; Thebault; Isoroy; La Boisserolle; Capelle; Museo Casa Mollino.**

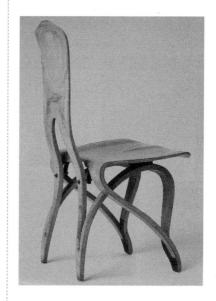

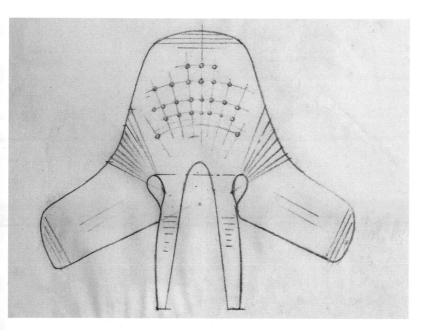

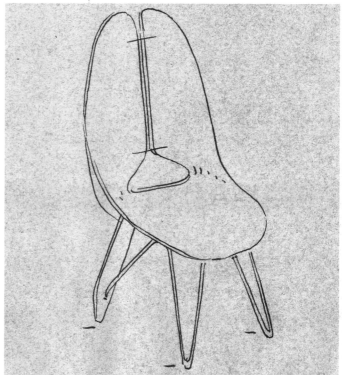

• *Left-hand page: Carlo Mollino's architecture. Pleated structures of the auditorium of the Teatro Regio, Turin, Italy, 1965-1973. (Photo S. Cavallo, courtesy Casa Mollino, Torino).*

• *Sketches of two laminated wood chairs for Casa Cattaneo, 1953. (Archives Carlo Mollino, Biblioteca Centrale di Architettura - Polytechnics of Turin, courtesy Casa Mollino, Turin).*

• *Chair and Tavolo Lattes, a laminated wood table for Casa Editrice Lattes, 1951. (Collection Bruno Bischofberger, Zurich, courtesy of Casa Mollino).*

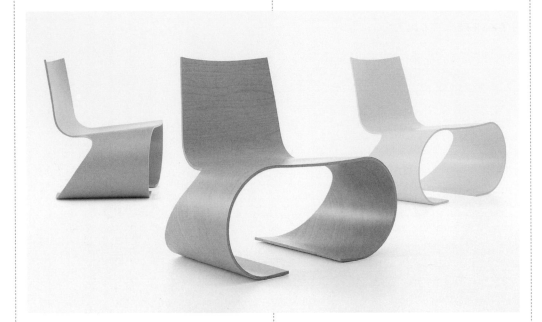

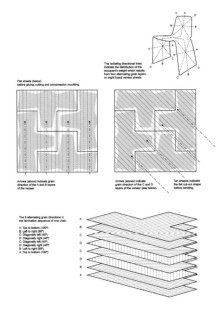

- *The curved shapes of Oto.*
- *The pop-up inspired folds of the Eco.*
- *Cutting pattern for the Nxt.*
- *Tri, stackable chairs.*

Voxia

PETER KARPF

Voxia is a collection of one-piece, laminated wood chairs created by the Danish architect Peter Karpf for Iform.

The chairs' originality lies in their patented design, comprised of successive multi-play veneer layers that are positioned alternately at 45° or 90° angles, which optimizes the chairs' strength and bendability.

An economical assembly process allows for the chairs to be cut from just one wooden panel. With their simplicity and the purity of their Nordic lines, this collection shows off the ecological and economical potential of moulded wood. The Xus model is moulded straight from a single piece of wood into a three-dimensional shape.

I Fold You

MARGRIET FOOLEN

"I Fold You' was my design project for my degree at the Design Academy Eindhoven. My stackable and weatherproof garden chairs are produced from a single sheet of curved beech plywood, and made using curves and cut-outs rather than screws or glue, making them strong and comfortable. I arrived at the design for my garden furniture through folding sketched paper models. You can see the lines of the folds in the finished products."

✎ **www.voxia.com;
www.margrietfoolen.com/en/info-en.html;
origami stools by Enrico Bona for Montina.**

• *By Margriet Foolen*

• *Nxt chairs by Peter Karpf.*

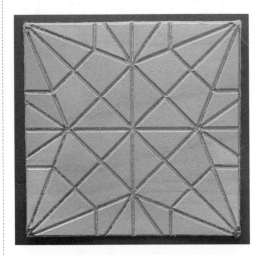

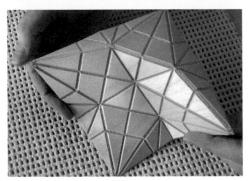

• *Sample piece of Foldtex showing its folding properties.*

Foldtex

TIMM HEROK

Designed by Timm Herok, Foldtex is a revolutionary type of lightweight and foldable board. It's a multi-ply material that comprises a 3mm base layer and at least one rigid and tear-resistant layer. When folds are marked out in the material by using a CNC* milling cutter a V-shape is created, with the flexible layer of the material acting like a hinge. In addition to being foldable, new surface layers can be added to it, such as linen, leather-based fabrics, PVC and reflective aluminium sheeting, meaning furniture created with the material can be customised with different textures and looks.

The production processes facilitated by Foldtex and the possibilities of combining it with other materials will undoubtedly prompt designers to re-think their use of traditional approaches such as corner connection and surface finishing techniques, which are rendered obsolete by Foldtex. Foldtex is made mostly from recyclable materials; basic Foldtex is made from between 60% to 80% plasterboard, 15% to 20% recycled leather, and 5% to 10% latex.. Combination of wallboard (3 or 1 mm) and one or two-sided coating (to 0.5mm), panel dimensions:
1 400 x 2 100 mm, from 1 300 g / m² to 2 200 g / m².

*CNC: Computer Numerical Control. Less complicated folds can be made using a hand milling cutter. Constructeur Treser-Georg-Str. 49, D-60599 Frankfurt, Germany.

✆ **www.foldtex.com**

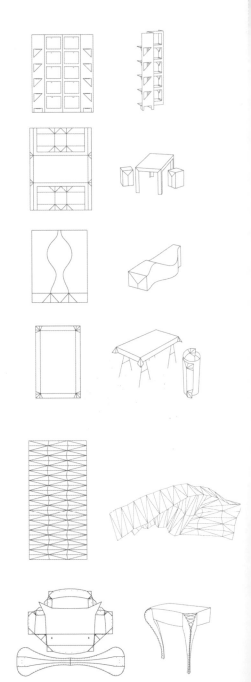

• Various creations by Tim Herok and Suzan D.
Cigirac: screen, 1999; tablecloth, Materialica Design
Award 2003; folding pattern and cutting template
for Sissi console table.

Wooden Screen

ANNE-LAURE CHEVALLIER

Whilst she was a student at the Ecole Pivaut in Nantes and living in a small studio, Anne-Laure Chevallier attempted to solve the main shortcoming of this type of property, namely the fact of a single room serving as a bedroom, a kitchen, a living room, and even an office.

"I thought about the pieces of furniture that were the most important and took up the most space in a typical student flat, arriving at the conclusion that the table and chairs should also be given the function of dividing up the space. So I thought about a screen that could be turned into an eating area with a table and two chairs. It seemed obvious that it would have to be easy and quick to deploy.

My goal was to create a piece of furniture that had a very simple appearance and was made out of a classic material such as wood, but which was nonetheless a design piece. The viability of the whole thing depended on its shape. It's made from 12mm to 18mm-thick glued panels of multi-ply birch, and wood hinges inside the body, made through cut-outs in the wood, so that the piece works as a partition when it is folded up. The support legs of the table and the two chairs are made of cut outs from the front panel and pivot outwards. Ball-bearing latches allow the three panels of the screen to move. By pulling back the panels the screen can be taken down."

Front view

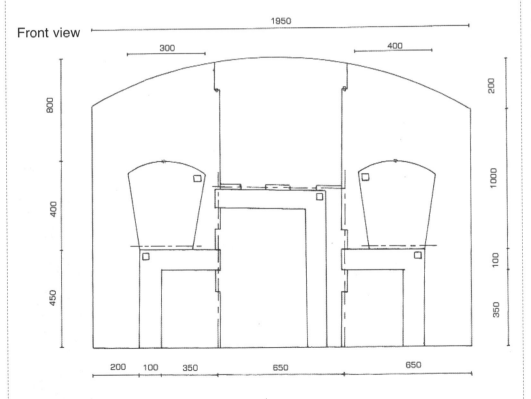

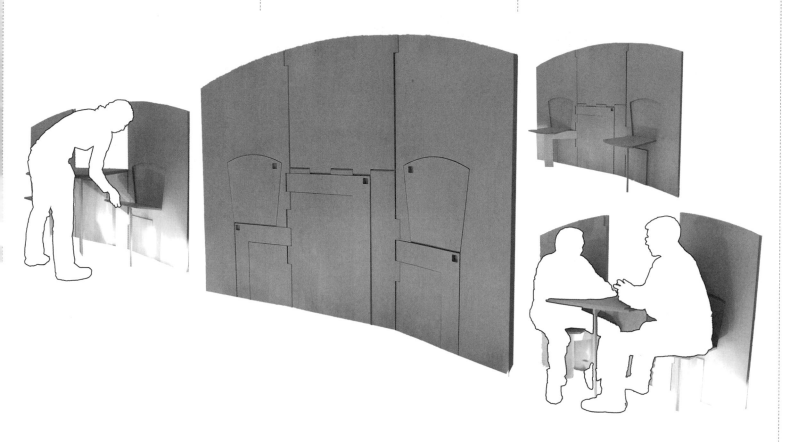

CARDBOARD FURNITURE

Generally speaking, the difference between paper and cardboard is their respective weights, with cardboard weighing over 200 grams per metre squared. Corrugated cardboard, meanwhile, is made from cellulose fibres that are 85% recycled cardboard fibres and 15% virgin fibres. It is made from outer layers of paper and layers of grooved paper that keep the different layers an equal distance apart from one another, meaning it works as a strut and a shock absorber (corrugated cardboard is used primarily in the packaging industry).

This type of cardboard varies in style from 2mm-thick micro-corrugations to grooves that are over 6mm wide, meaning it is a multi-purpose material. In terms of thicknesses, corrugated cardboard comes in simple face, double-double face and triple corrugated.

MARKING OUT FOLDS

Scoring is the process through which cardboard is weakened by crushing the cardboard fibres at the fold points to make the folding process easier and prevent cracks developing (as with Bristol paper, for example). In the paper industry, a scoring tool is generally used, made from a rounded steel blade and a precision-cut guttering made from a hard material. The blade pushes the card into the guttering placed under the car to create a groove. The process can also be carried out using smaller, hand-powered tools or a scoring machine.

• Won, *blind, four folding sections, 60 x 176 cm.*
Designed by O. Leblois, 1994. Decoration by Kicilievitz.
• T.4.1 armchair, *Designed by O. Leblois, 1993.*

CARDBOARD FURNITURE

This simple material, synonymous with fragility and a limited life span, can nevertheless – and in spite of appearances – be used to create some astonishing furniture. Creating such out-of-the-ordinary pieces requires careful thought and a thorough analysis of forms and forces.

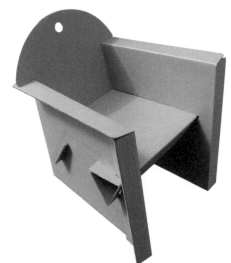

T.4.1.
(Tea For One)

OLIVIER LEBLOIS

An architect and lecturer at the Ecole spéciale d'architecture in Paris, Olivier Lebois is the designer of the fêted "T.4.1" (Tea For One) armchair. He has created some very intelligent, and simple yet aesthetically pleasing pieces of furniture from cardboard, including chairs, desks and partitions.

In the view of Isabelle Millet, the director of the Quart de Poil gallery in Paris – which has been selling Leblois's furniture since 1993 – "the cardboard chair has become part of the permanent collections of many different museums. For the most part the cardboard is sourced from sustainably managed forests. Solid board or single- or double-corrugated cardboard, for example, are made from 80% to 100% recycled materials and water-based adhesives. The assembly of this furniture is a simple process which doesn't require tools or glue, and instead just requires slotting the pre-folded cardboard together, meaning the pieces can be re-used if they are disassembled. The edges produced in the cutting process are also smoothed, meaning the chairs are also very safe."

T.4.1 armchair: 71 × 64 cm, seat height 35 cm.

🖷 **www.quartdepoil.fr**

• Assembly process for the T.4.1 *armchair*.
• Lotus *chairs*, various decorative styles. Made from a pre-folded piece of double-corrugated cardboard, with a reversible and detachable back and locking key. Designed by Stéphane Millet architecte, known as Essaime, 1995.
• Lotus *chair*, decorative design by Annabelle Girard.

• Dondo, *chaise longue coated in wood or with a fabric covering*
• Poltrona Lady.
• Lady, *armchair made from double corrugated cardboard.*

Dondo

NOMADEDESIGN

Dondo is a very clever rocking chaise longue. The brainchild of Italian designer Generoso Parmigiani, Dondo is made entirely from double corrugated cardboard and was produced in a limited run of 101 chairs, each one numbered and signed. Thanks to its specially curved design, the chair can be manoeuvred simply by either pushing one's arms forwards (to put it in a sitting position) or pulling them back (to make it recline). The chaise longue's unique design and its high quality and intelligently used cardboard make it capable of bearing a weight load of more than 350kg.
If desired the chair can also be covered with a fine layer of laminate, multi-ply wood or felt made from recycled fibres (L: 1,70 m; W: 55 cm; H: 75 cm.)

✏ **www.nomadedesign.com**

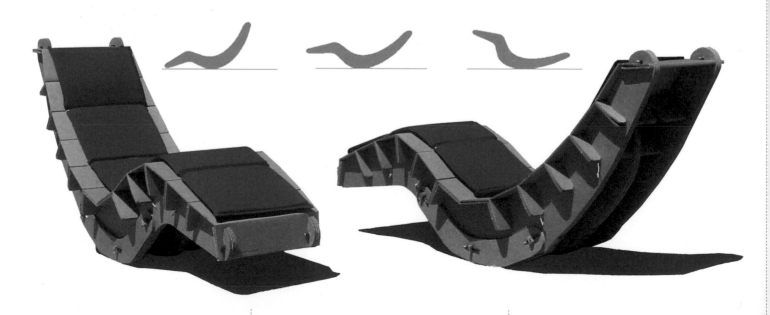

• *Hydraulic cutter at the Nomade Design workshop.*

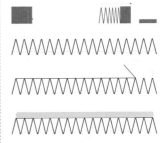

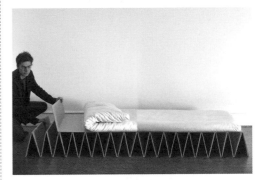

Itbed

VALÉRIE JOMINI AND STANISLAS ZIMMERMANN

Whilst students, Zurich-based architects Valérie Jomini and Stanislas Zimmerman often found themselves moving house. This situation gave them the idea of creating a cardboard bed that could fold up like an accordion. The weight of the mattress is supported by several supports, giving it a high level of stability. There are industrially-produced folds every 25cm in the 7mm-thick corrugated cardboard of the Itbed. The bed is assembled with adhesive tape, and is given stability by two lengths of webbing that run from the head to the foot of the bed. Looked at from the side, the Itbed bears a certain resemblance to a house of cards. Length: 90 and 140 cm. Creation date: 1997.

✏ **www.it-happens.ch**

• *Demonstrating the easy transportation and setup of the Itbed.*

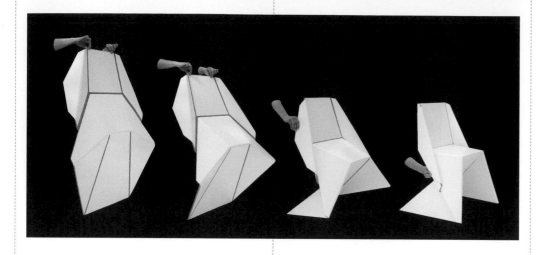

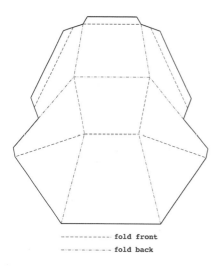

-------------- fold front
------·-------- fold back

• *The* Papton *chair is very light, and as easy to fold as a paper aeroplane.*

Papton Chair

WILM FUCHS AND KAI FUNKE

The Papton Chair is the work of German designers William Fuchs and Kai Funke, who with just a few folds transformed a sheet of honeycomb cardboard into a very light-weight chair (2.4kg). The structure of the chair is based on separated panels that divide up the load-bearing and the flexing zones of the chair. The chair's lightness and compactness — up to 80 chairs will fit into a standard palette — make it ideal for meeting spaces. Its distinctive shape, made up of pairs of polygons, is based on a simple folding pattern that gives the chair a sense of lightness. Disassembled, the cardboard sheet of the chair measures 1335 × 1180 × 10mm. When assembled, the chair has the following dimensions: H: 800mm, W: 620mm, D: 530mm
(FUCHS + FUNKE Industrial Design, Berlin.)

Foldschool

NICOLA ENRICO STÄUBLI

Bern-based Swiss architect and designer Nicola Enrico Stäubli offers us the concept of 'foldschool', a collection of build-it-yourself cardboard furniture for children. The plans, templates and instructions can be downloaded from Nicola's website.

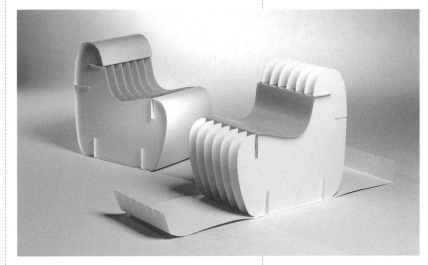

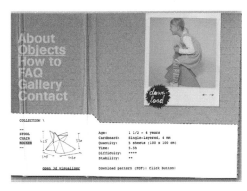

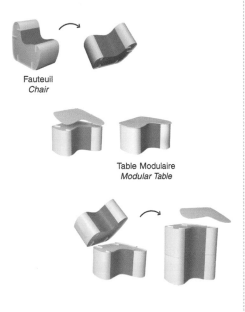

Fauteuil
Chair

Table Modulaire
Modular Table

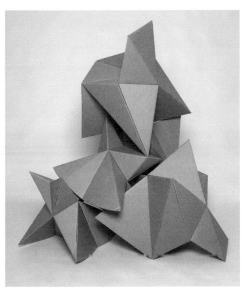

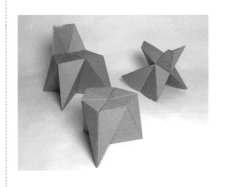

AdSeat

GEORGES CAUX

The AdSeat range of modular furniture is made from both flexible and rigid polypropylene panels that are cleverly assembled without the use of screws or glue. Caux's modular design concept is very widely used at exhibitions due to the ease with which the pieces can be transported and installed. User customisable, this furniture can be transformed into a sofa, counter, coffee table and many more besides.

Corner-Bar   **Canapé**
Sofa

• AdSeat furniture base module.

✏ **www.ondef.org; www.cofepac.org; www.fuchs-funke.de; www.raacke.com/; www.foldschool.com; www.adseat.com; http://www.okupakit.com/index.htm; www.bluemarmalade.co.uk/en/products-lbpop.htm; "Folder" polypropylene chair by Stefan Schöning for Polyline.**

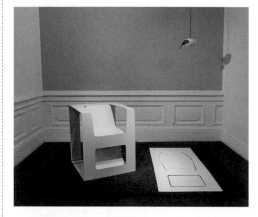

Slimmy Chair

Designer Frederic Debackere develops his pieces based around a philosophy of easy transportation and combining folds with internal spaces. His minimalist, cube-shaped and pop-up inspired Slimmy chair has a refined aesthetic and weighs only 4.5kg. Slimmy is produced from a single sheet of a composite material of PVC and aluminium, which provides durability for both indoor and outdoor use.
The material itself serves as the hinges of the piece. The panels of the chair reinforce one another in a way that means that no other reinforcing materials are required. The chair is a good demonstration of the properties of the material from which it is made. The chair is delivered flat and is easy to unfold, with the assembly process requiring just four screws. The chair can be assembled with the seat at either a right angle or a more inclined 100°.

Lounge Chair

ENRICO WILDE AND MEHRWERK DESIGNLABOR

Enrico Wilde and young German designers Mehrwerk Designlabor envisaged producing, out of a sheet of sandwich material with a type of honeycomb cardboard inner layer, an original piece of furniture that would be "a light and foldable lounge chair made from recycled materials. The originality of the AufJedenFalz lies in its use of origami folding, which allows it to be made from a flat, rectangular sheet and minimises wasting primary materials during the production process. The exterior layer of the chairs is made from natural fibres (in this case silk), while the internal layer of the sheet is made from honey-comb cardboard. A carbon-neutral linseed oil epoxy resin is the other raw material we use. Using these materials, we were able to produce a prototype that weighed just 5.5kg, but which was capable of holding a weight of 120kg. Adding textiles to the composite material helped to strengthen it. In future, we want to develop three-dimensional textiles designed especially for furniture production. This would be a major step forward in the history of textile design, as ever since they have been produced indus-trially textiles have always been thought of as a two-dimensional material manufactured by the metre."

✏ **www.borndesign.net ; www.mehrwerkdesignlabor.de**

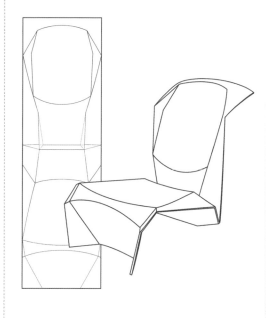

• AufJedenFalz, Lounge chair, floral pattern, designed by Enrico Wilde and the Mehrwerk Designlabor, 2006.
• Folding pattern from a rectangular sheet, avoiding material wastage.
• AufJedenFalz lounge chair and, close-up, honeycomb structure and hemp fibres.

Cutting-Edge

THOMAS DUMOULIN AND NICOLAS MARZOUANLIAN

The clean lines of this seat-cum-sculpture were designed by Thomas Dumoulin and Nicolas Marzouanlian. "The first chair we created for the Cutting-Edge project was based on a simple idea: making robust and light pieces of furniture that could be used both inside and outdoors. As we worked on the project its form, initially just created out a piece of paper, took shape and became more refined as a result of creating several models that clearly demonstrated that using folds in the structure would both make it rigid and open up avenues such as giving the piece an abstract sculptural dimension. However, the more the project advanced, the bigger the problem of how to execute it became. Sheet metal, the material we initially used, no longer worked with the designs we had made. It was during a discussion with an associate that a solution came to light: aluminium composite panels. Light and durable, and also easy to manufacture and foldable, they fitted the bill perfectly."

✏ **http://cutting-edge.fr/**

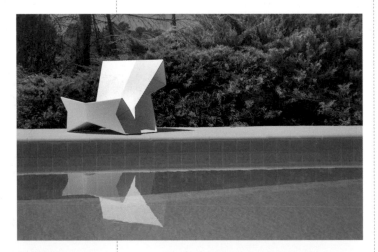

• FL *model.*

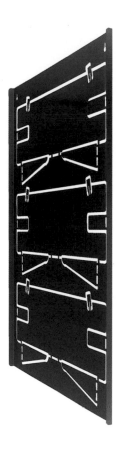
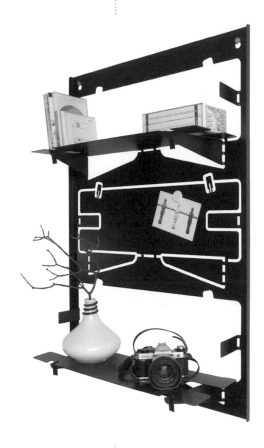

Piegato

MATTHIAS RIES

Delivered flat and made from just one piece of sheet metal, the Piegato has a high load-bearing capacity. Its designer, Matthias Ries, arrived at this design, which results in very little waste materials, as he attempted to make savings on production costs. Laser cut and painted, the Piegaro can be transported very easily. The shelves on this new breed of storage unit fold out by hand, and the whole piece can be installed in just a couple of minutes and only needs two screws.

Dimensions: height 100 cm x 66 cm x 2 cm; weight: approx. 8 kg.

✎ **www.mrdoproducts.com**

1

2

3

4

5

6

7

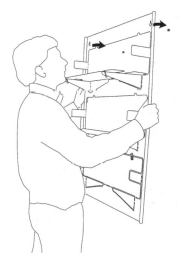

From Flat to Folded

RAQUEL VEGA

Swiss designer Raquel Vega's approach to creating is "from flat to folded, from folded to origami". Vega had the idea of taking materials found on construction sites, especially sheet metal, and treating them like sheets of paper, creating forms through folding. Her armchairs subtly combine treated, rusted sheet metal with stainless steel. Her current work revolves around the principle of going from "origami to articulated folds".

"Folding is above all a response to problems of equilibrium and strength, and then a form of artistic expression or a state of mind or philosophy. Whilst involved in construction work at the Shakan art gallery in Lausanne, I found myself with 10 surplus sheets of rusted metal. They had been used to cover the floor, and I didn't know what to do with them as they buckled and bent in all directions under their own weight. But then I had the idea of making them rigid through folding them. I took the sheets to the workshop of a friend of mine who is a locksmith, and we machine folded and then cut them. I took the (1m x2m) plates we made home, and used them to design Fauteuil 1 and Fauteuil 2 for the Shakan gallery. I went on to use the same technique in the rest of my creations.

My goal in designing is to give 'stability' to the 'sheets' through folding, and to create form and shape. My intention is to create functionality in a simple and elegant way, using 'sheets' to create form in three dimensions. Applying these ideas to architectural elements such as staircases is also possible, and doing so creates volumetric sculptures.

Experimentation is of crucial importance; we need to continually evolve and visualise new possibilities. Doing so keeps us engaged and opens up our eyes to new things; transforming and re-appropriating materials to constantly save money and time becomes almost a reflex response. When it comes to questions of costs and efficiency, my motivation has always been to provide the best possible answer by re-using industrial materials and 'waste'."

REMARKS FROM MARCH 2007.

🖙 **www.designvega.ch**

• Fauteuil 1, *materials: rusted metal and stainless steel, 70 x 70 x 70 cm, weight: 15 kg.*
• Tupperware1 *stool.*

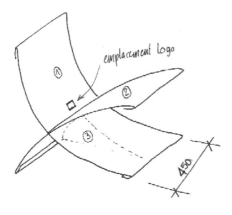

emplacement logo

450

finition inox le plus fin possible
micro-billé

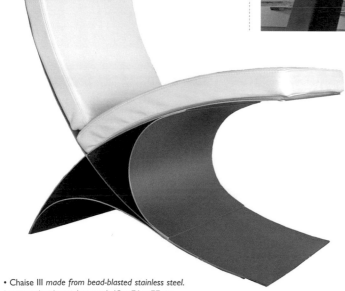

• Chaise III *made from bead-blasted stainless steel.
Laminated with stainless steel, 45 x 76 x 77 cm,
weight: 20 kg.*
• *Renovation of a house in Grandvaux, staircase made
from folded sheet metal (made in collaboration with
Lausanne-based architect M. Luis Crausaz).*

Bent

STÉPHAN DIEZ
AND CHRISTOPHE
DE LA FONTAINE

Designers Stéphan Diez and Christophe de la Fontaine use 3mm-thick, laser-cut, pressed and folded aluminium sheets to make the pieces in their Bent collection, which was presented at the Ideal House exhibition in Cologne in 2006. The range includes a sofa, armchair and stool. The aesthetic perforations that mark out the folds of the furniture are the hallmark of this collection. Perforating the material allowed it to be folded in a single stage, unlike the several stages required in more traditional processes.

Using a similar technique of working with perforated metal, only in the field of design objects, Astrid Traissac created "Design et Dependance" after meeting with metallurgists and designers. Traissac works with metal sheets as though they were flat pieces of fabric, cutting from templates, making folds and assembling to create innovative objects with original aesthetics.

Pichic

ENNIO VICENZONI
AND INGE DE JAGER.

This novel piece of furniture, ideal for outdoor use, is made from a single sheet of aluminium, meaning the benches and table form a single whole. The piece is decorated with a cut-out floral pattern.
Table: 156 x 190 x 80 cm.

✎ www.vicenzoni-dejager.nl

OTHER MATERIALS

Some designers do not hesitate to use new materials and play with their properties (for example, malleability, coarseness or sheen). Sascha Akkerman and her 'Poissonmobile' sun lounger is one example. In a similar vein, in addition to creating a variety of tables and chairs designed out of glass or aluminium folded in a minimalist style, Cecilia Lundgren designed "Vita", a coffee table made from folded acrylic. Daniel Michalik, meanwhile, has made use of cork to create a sensuous chaise longue.
And last but not least are Sooin Kim's Cardine chair, created from a sheet of plastic held in position by Velcro, and Jake Phipps's astonishing Isis chair.

✎ www.moroso.it; www.stefan-diez.com
www.jakephipps.com; www.danielmichalik.com
www.ceciliadesign.se/vita.html
www.indexaward.dk; www.confused-direction.de
www.sabz.fr/fiche-produit/ guggenbichler/ banc
www.designdependence.com/fr

WALL FURNITURE

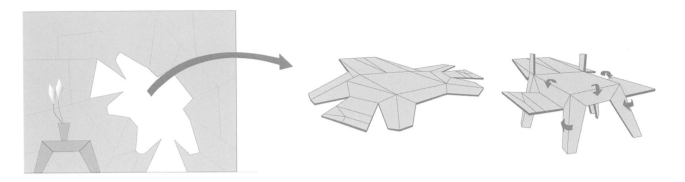

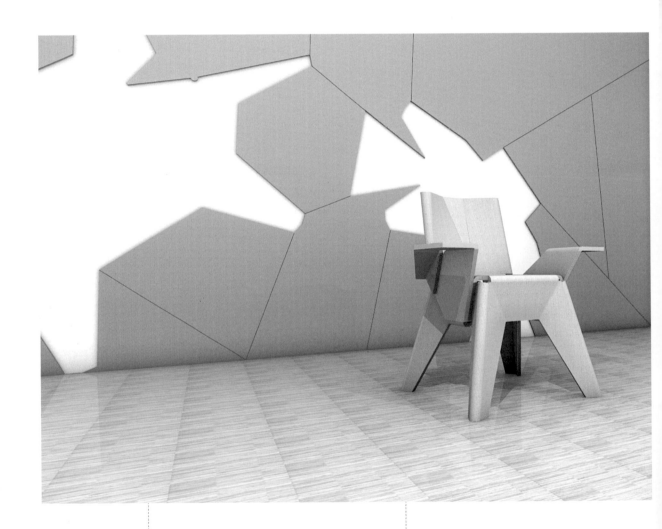

Folding Chair

PHILIPPE DUFOUR FERONCE AND LISAMARIE VILLEGAS AMBIA

The idea of freeing up floor space when furniture is not in use by turning it into wall décor is one that designers have occasionally turned their attentions to.

For their entry to the Photo Real Future 2005 International contest, which focused on furniture and photorealism, designers Philippe Dufour Feronce and Lisamarie Villegas Ambia made use of multi-layered felt to create an innovative and modern fresco.

"The idea behind our foldable chair made from felt was to turn your living room's wallpaper into furniture. Large felt pieces are placed on the walls, and some of them can be detached to create furniture. Felt is a very rigid, warm and cosy material. Velcro-type fasteners are used to both keep the felt pieces on the walls and to assemble the furniture."

There are two different types of felt: felt made from wool, the fibres of which adhere to one another naturally – this is the classic and natural type of felt – and synthetic needled felt, made from various types of fibres and resins

✆ **foldable wall chair by Dror Benshetrit: www.studiodror.com/; www.lefeutre.fr/**

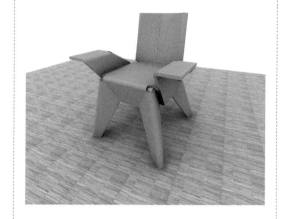

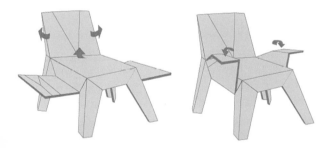

Pouf

GREGORY LACOUA

This project was created by Gregory Lacoua during a project workshop at the ENSCI (Ecole nationale supérieure de création industrielle, also known as "Les Ateliers"), and funded through the VIA (Valorisation de l'Innovation dans l'Ameublement) This original and eloquent pouf stool can transform, as if by magic, into a comfortable rug. Its structure is made up of beech plywood panels overlaid with foam. Pulling on the button in the middle transforms the piece via the laces that connect the jointed panels.
Pouf : 75 × 75 cm, h : 44 cm and, as a rug, 120 × 120 cm.
Ligne Roset - Label VIA 2007.

✇ **www.ligne-roset.tm.fr**

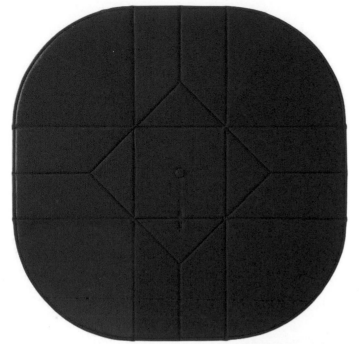

Fanfan

HIROKI TAKADA

Hiroki Takada's originality as a furniture designer is expressed through a range of materials. His Fanfan chair (2002) makes use of stainless steel, aluminium, felt, polyurethane sponge and wood.
Dimensions: large model, 165 cm × 65 cm, h: 96 cm.
After seeing Botticelli's The Birth of Venus at the Uffizi in Florence, Takada came up with his shell-shaped, plastic and aluminium Venus chair.
Dimensions : 80 cm × 85 cm, h : 85 cm.
In 2007 he created a tall stool, inspired by The Winged Victory of Samothrace (currently held by the Louvre), with wing-like armrests.
Dimensions : 150 cm × 80 cm, h : 140 cm.
Plastic, stainless steel.

✆ www.takadadesign.com

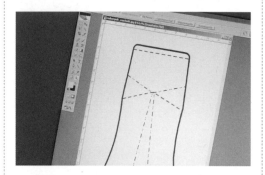

Cut'n Fold

Hannecke is a company that has been mastering the art of working with Plexiglas for the last thirty years. It has developed Cut'n Fold, a three-dimensional shaping technique for Plexiglas that has potential new creative applications in the fields of design and architecture. Creating simple bends in plastic using thermoforming (reshaping the material using heat) is a common practice, but successfully creating a curved shape is more complex.

The technique that has been developed fixes the shortcomings of the more traditional methods of bevelling or thermoforming, which alter the properties or the transparency of the material. Dr Stefan Delecat, an origamist who works for the company, was closely involved with the development of this process. He has carried out experiments with putting folds in Plexiglas for several years, especially in the area of creating curves. The pieces are pre-cut, and only the fold lines are heated. All other parts are kept cold and adapt to the new shape during the shaping process. This improves the piece's distribution of bend points and tensile strength. The low costs of the process make it possible to quickly produce small lines of display shelves and cases, light fittings and furniture.

✏ **www.cutn'fold.de**

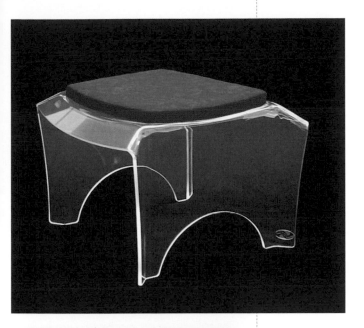

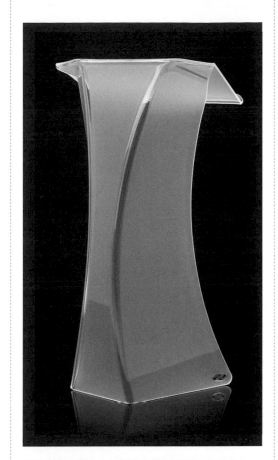

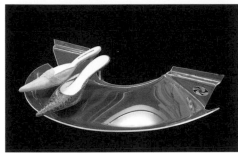

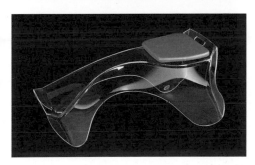

COMPOSITE AND GLASS FURNITURE

Panels made from composite materials – sheets assembled out of layers from different families of materials – are becoming increasingly important in the world of furniture design.

The best-known composite panels are:

- ALUCOBOND, made from two 0.5mm-thick aluminium sheets and a polyethylene core;
- TECU BOND, made from two 0.5mm-thick copper sheets and a polyethylene core;
- DIBOND, made from two 0.3mm-thick aluminium sheets surrounding a high-density polyethylene core.

DIBOND is flexible and can be bent to a small radius without heating. The Wide Chair, created in 2004 by Aleksi Penttilä from the Finnish Rehti collective, is a good example of the applications of DIBOND.

T-Vertigo: "the swinging table"

AQUILIALBERG

Designed in 2007 by AquiliAlberg for Moroso and made from a strengthened polyester resin with a lacquered finish, the T-Vertigo table's design creates a visual effect of rotation and the illusion of continually flowing lines. Ergian Alberg and Laura Aquili's Ecsher-inspired table transmits an original sensory quality through exchanges between solids and empty spaces and between lightness and dynamism. The expressive sinuousness of the design generates a perception of movement in an everyday piece of furniture.

Dimensions: 90 cm x 90 cm, h: 27 cm.

✏ **www.aquilialberg.com**

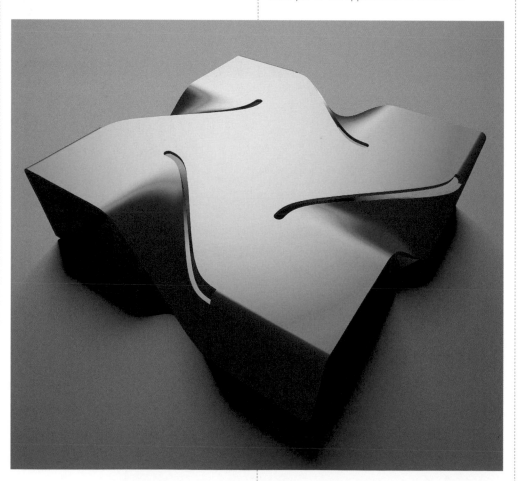

Ghost, Momento

CINI BOERI

In 1972 Vittorio Livi, the founder of Fiam, created his first pieces of furniture from single sheets of kiln-formed glass. The most common of the two existing techniques for shaping glass uses ceramic-fibre moulds that are resistant to high temperatures. Mireille Mazet, an art historian who specialises in glass-working, provides the following outline of kiln-formed glass working: "Kiln-formed glass-making techniques have been around since Antiquity. Producing kiln-formed glass entails moulding glass – generally flat sheets of the material – by softening it in a kiln while it is placed over a shaped or hollow mould. We distinguish between kiln-forming to make solid three-dimensional shapes and kiln-formed glass working achieved through the forces of gravity, a process frequently used in contemporary glass-making in conjunction with other techniques, especially fusing. The gravity-based process is known as slumping, and involves placing a sheet of glass on a pre-prepared mould, with the glass taking on the shape of the mould. The temperature in the kiln is steadily raised to 850°C or 900°C, before suddenly being turned down to 600°C. Gravity-induced free-fall slumping involves heating and softening the piece of glass over a hollow support. When the temperature reaches around 700°C the effect of gravity will make the glass change shape, stretching until the desired effect is achieved. This is followed by a period of reheating and then a cooling period. All types of flat glass can be kiln-formed, but they cannot be mixed together."

✆ **Pôle Verrier: www.idverre.net**
www.fiamitalia.it

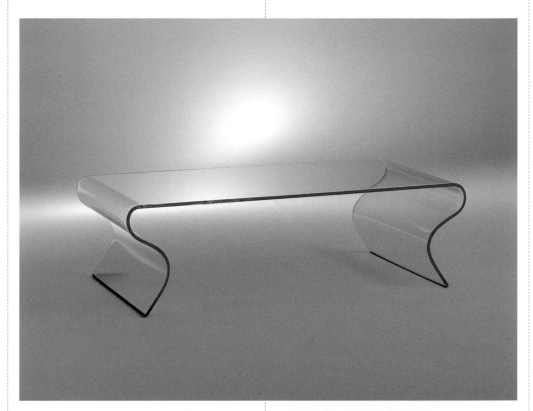

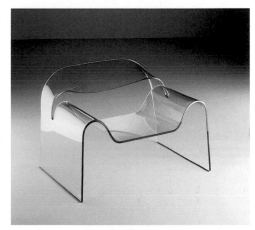

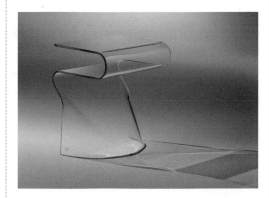

• *Charlotte, designed by Prospero Rasulo, Fiam.*
• *Grillo, designed by Vittorio Livi , Fiam.*
• *Ghost, designed Cini Boeri, 12mm curved glass, Fiam.*

• MAD *tables and chair.*

Ductal®: Creative Concrete

In recent years, people have started to talk about 'concretes' instead of simply 'concrete', a change in terminology that recognises the significant number of different possible compositions of the material. These varieties include a creative type of concrete called Ductal.

There are different members of the Ductal family, with distinctions being drawn based on their respective compositions, which in some cases include organic or metal fibres. The form of interest to us here is Ductal®-FO, which was formulated for making street furniture, panels and architectonic shells. It is an ultra high-performance, fibrous concrete (UHPC), made from reactive powders, additives and organic fibres. It is prepared using an industrial concrete mixer, and the resultant mixture is then poured into moulds or formwork. This type of concrete can be polished or sanded to give it different textures, such as a stone effect. The fineness of the concrete's grains and the fluidity of its composite material make it highly adaptable. Because of its unique characteristics, the formwork into which it is poured needs to be carefully and precisely designed. But the real significance of Ductal®-FO lies in the fact it can be used to make large-scale yet very thin (8-10mm) structures without the need for passive reinforcements or metal frameworks. When making folded or curved structures the only factor that needs to be taken into account when producing the bend radius is the thinness of the formwork. In addition to being highly resistant to shocks and compression, not to mention bending – depending on the concrete's fibre content – the material can also be made less porous through heat treatment.

Examples of Ductal in use include the flower pots and guard railings of the Flower Tower in Paris, the benches and garden boxes of the Place de la République in Rennes, and the acoustic panelling of the TGV station in Monaco.

The Mad range, designed by Jérémy Bataillou and produced by Atelier du béton, comprises outdoor furniture, including an armchair, a bench, a coffee table and a foot rest, designed using simple and clean lines that demonstrate this material's capacity to be used for creating very thin layers.

✏ **www.atelierbeton.com; Francesco Passaniti, www.compactconcrete.com; Milène Guermont, concrete furniture**

264

720

720

720

50 kg

720

724

374

716

62 kg

Alphabet

MARIE COMPAGNON

Marie Compagnon has a creative approach all of her own. Jeanne Quéheillard, a lecturer at the Ecole des Beaux-Arts in Bordeaux, sums it up in the following terms: "Out of the flat of a 2D pattern, she creates her articulated architecture 'Alphabet' of metamorphosing objects and spaces. Screens, partitions, shelters, rocks and hideouts can all be freely constructed and deconstructed from a single, saddle-stitched sheet (made from wood, foam and felt). Her 'Alphabet' takes on forms that are held in place through pin hinges, like the ones found in windows or shutters, in this case made from metal rods and leather rings. These things have not come about by chance, and they make their purpose clear: inhabiting the objects, and stepping on them or letting them envelop you. There is little difference, then, between the piece's construction method and manipulating it to make objects. The foldable shapes are her 'workable scenery'. The mobility of the joints is grounded in familiar techniques such as sewing and punching. She stubbornly accepts them, in spite of the possibilities of using more industrial methods. What one experiences is the programmed mechanics of the movements, which come from its construction of friezes, clusters and intrinsic motifs."

Alphabet: little articulated architecture

"Initially designed as an intricately shaped felt rug, Alphabet is a proposal to redefine the notion of space. Its articulated design allows it to evolve into three-dimensional shapes, to become a screen, partition, shelter, rock or a hideout. It can be freely constructed and deconstructed, constantly changing form and function to either be trod on or lived under. My creation process was to first of all make a 3D computer model, sculpted like a rock. I then 'unfolded' it to get a flat pattern, which is the rug. Each fold is a hinge, allowing the piece to be freely folded, unfolded and refolded."
REMARKS FROM JANUARY 2008.

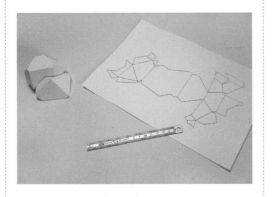

• Alphabet, little articulated architecture, model viewed from above. Cardboard, felt and cotton yarn, 2004.

• Alphabet, little articulated architecture.
Prototype made by Jacques Gauthier, saddler.
Felt, plywood, cotton yarn, metal and leather,
2005.

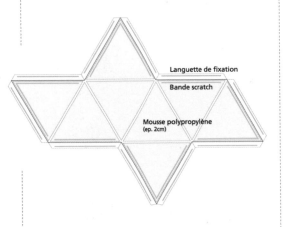

Languette de fixation

Bande scratch

Mousse polypropylène
(ep. 2cm)

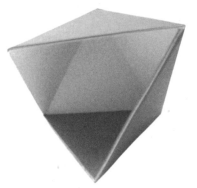

Siesta Origami

HELMINA SLADEK
AND PAULINE THIERRY

Helmina Sladek and Pauline Thierry, architects at the Ecole d'Architecture in Strasbourg, dreamt up their 'origami à sieste' ('siesta origami') concept for their entry to the biannual Minimaousse 2 microarchitecture contest of March 2006, organized by the la Cité de l'architecture et du patrimoine museum in Paris.

The contest took place in two phases. Firstly, there was a call for entries, followed by a final stage in which shortlisted entrants were invited to Isle-d'Abeau to produce full-scale versions of their designs. The goal of the competition was to demonstrate, through competing entries showing off their respective inventiveness and artistry, the role that small-scale projects can play in architecture. This project was included in the Institut Français d'Architecture's 'coup de coeur' award category.

"We were looking for a stable and coherent starting point in response to the question of creating transportable micro-architecture. Origami seemed like an obvious solution, and the expression 'plier bagages' ('pack up and go', or literally 'fold up your baggage') took on a very real dimension to us.

However, we took a step away from one of the key precepts of origami by folding a non-square shape. Under our approach, origami allowed us to create a design that was very easy to assemble and use.

The idea behind the project is to escape from the city, its noise and stress, and head off to open spaces with a practical shelter to relax in and give you some privacy; or being able, in just a few moments, to take this light and compact shelter with you on your bike and set it up quickly. Origami doesn't need any accessories: it is self-sufficient. The panels of the shelter fasten to one another with strips of Velcro. Made up of eight 1.5m wide equilateral foam triangles, it also works as a reversible rug or

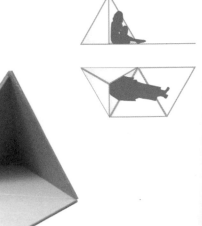

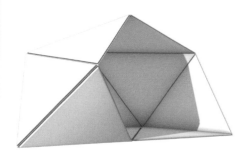

mattress. It's a light, re-shapeable piece of origami. You give it the shape you want, with the panels folding either way with equal ease. Two pieces can be linked and assembled together, creating even more folding combinations. And why not combine several together at once? It's up to you to experiment!"

DEPLOYABLE STRUCTURES

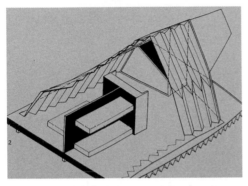

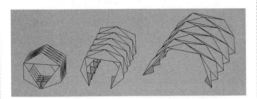

Perhaps the best known and quintessential deployable structure is the yurt and its ingenious jointed wooden skeleton. Designed to be set up quickly and to be easily transportable, the yurt is a good example of 'unfoldability'.

But objects can also be deployed through other means, such as stretching, extending, swinging, rotation, twisting or turning.

There are many examples of the deployable, such as Herbert Yates's experimentation with simple and classic accordion folds, the Plydome shelters used by migrant farmers in California, designed and produced by Sanford Hirshen and Sim Van der Ryn between 1965 and 1974, or Guy Rottier's 1983 extendable vacation home. More recently, the Brazilian-designed dB.Folding Disco, made from coloured PVC panels, Joerg Student's Ha-ori Shelter for homeless people and Carolina Pino's Shellhouse provide answers to similar questions.

Some systems combine a structure or skeleton with textile membranes that are kept taut by the framework. There are too many architects and engineers who have taken an interest in these types of structure to mention them all, but amongst the most notable are the creations of the architect Renzo Piano, especially his mobile sulphur extraction factory made in Pomezia, Italy, in 1966. The English architect Arthur Quarmby investigated this area during the 1970s, and since 1990 Sergio Pellegrino's teams at the Department of Engineering of the University of Cambridge have also been carrying out research into it. In 2001, led by Prof Richard Liew, the Department of Engineering of the University of Singapore investigated creating an umbrella structure out of a textile membrane.

Another landmark piece is the Pinecone. Based on Ian Stewart's research into deployable structures, it makes use of applying mathematics to origami. Maths fans are strongly advised to look into the work of Erik Demaine, a computer scientist and professor at MIT, on deployable forms and their industrial applications, as well as older work carried out by Ron Resch, a mathematician and pioneer of origami structures.

• *Plydome, Sanford Hirshen and Sim Van der Ryn, architects.*
• *Research into folded structures, J.M. Prada, architect, in the* Pabellon de Venezuela, *expo'92 Seville.*
• *Jumbo origamic arch white, Bow Wow Architects, Jakarta, 2005.*
• *The archetypical deployable structure, the walls of a yurt are constructed from several wooden lattices whose joints are created by 'pins' made from tied leather.*

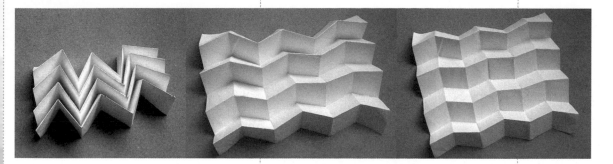

The Miura Fold

Travelling through space propelled by the sun's rays is no longer a fantasy! Olivier Boisard of the Union pour la promotion de la propulsion photonique says that: "Halfway between a present-day dream and tomorrow's reality, solar sails are now on the starting line for launching us into space."

In 1980 studies into using sail-like technology in space began, with the U3P (Union pour la promotion de la propulsion photonique) beginning to develop its vision for solar sails. To learn about the skills of sail folding, the U3P called on the assistance of the MFPP (Mouvement français des plieurs de papier), who were able to provide a solution. An alternative model comes from the system invented by the Japanese engineer Koryo Miura, which allows a sheet of paper (or indeed a road map) to be easily unfolded and refolded just by pulling or pushing the two opposite ends of the material. Miura's folding pattern is used nowadays in Tokyo tourist maps.

• Research into deployable structures: D.D. Focatiis and S. Guest.

Leaf Structures

University of Cambridge research scientists Davide De Focatiis and Simon Guest have been involved in analysing the structures of leaves, and in particular those of the Carpinus Betulus (hornbeam) and the Fagus Sylvatica (beech). Their research provided them with a better understanding of the leaves' natural properties and in turn allowed them to propose ways they could be applied to deployable structures or thin membranes such as solar panels, solar sails, tents or mobile shelters. One of the features of leaves is that they are a good middle ground between flexibility and rigidity, being able to support both their own weight and other forces or objects such as wind or snow. It was this characteristic that prompted research into the positioning of their veins. De Focatiis and Guest's mathematical research demonstrated that the simple model of regular and alternating valley and crest folds found in the leaves of a hornbeam provide a structure that is ideal for deployable structures.

• *Model of the holiday home; the edge folds provide the exterior cardboard walls with greater stability. Guy Rottier*

Cardboard Holiday Home

GUY ROTTIER

Guy Rottier, an imaginative and visionary architect and a man far ahead of his time, created his "Maison de vacances en carton" in 1968, in an attempt to promote temporary forms of architecture and urbanism, which he considered to be of pressing importance. Intended to be delivered as a kit, and with a limited life of around three months, his design could be erected in just a couple of hours. The home would be made from basic materials, namely 10mm thick packaging cardboard for the walls and a plastic film held in placed by cardboard poles for the roofing. Unfortunately, the home never became commercially available.

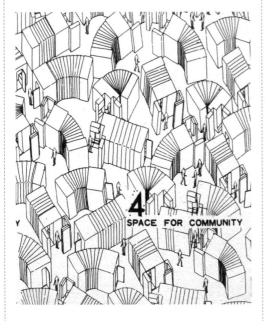

Nihon University Project

MASAYUKI KUROKAWA

In 1971, Masayuki Kurokawa proposed an end-of-year project to his students at Nihon University based around using f urnishings in an architectural design. The challenge delivered surprising results. The students' work led to an architectural arrangement comprised of an array of standardised (identical) and fixed elements that were connected together by movable accordion-folded coverings.

• *Variable geometry holiday home, Guy Rottier, architect, 1983.*
• *Facing page: Holiday home made from flat, cut 10mm packaging cardboard, 1968.*

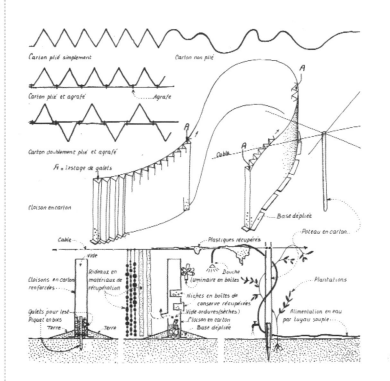

Carton plié simplement   Carton non plié

Carton plié et agrafé   Agrafe

Carton doublement plié et agrafé

Fi = lestage de galets

Cloison en carton

Câble   A

Câble

Base dépliée

Poteau en carton

Plastiques récupérés

vide

Cloisons en carton renforcées

Rideaux en matériaux de récupération

Douche
Luminaire en boîtes

Galets pour lest
Piquet en bois   Terre   Terre

Niches en boîtes de conserve récupérées
Vide-ordures (sèches)
Cloison en carton
Base dépliée

Plantations

Alimentation en eau par tuyau souple

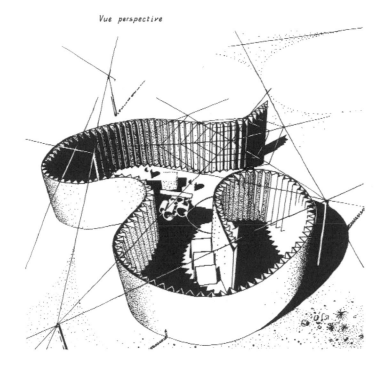

Vue perspective

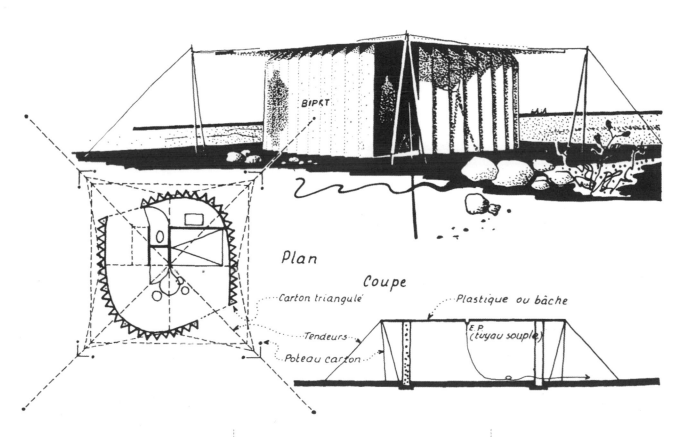

BIPKT

Plan

Coupe

Carton triangulé

Plastique ou bâche

Tendeurs

E.P. (tuyau souple)

Poteau carton

CARDBOARD STRUCTURES

Cardboard has many different applications beyond its traditional use as a packaging material. A 1905 edition of the Revue des Sciences (held at the Conservatoire Numérique des Arts et Métiers) reveals that G. Lougreux had already put the material to use as a building material at this time. In this case, waterproof cardboard, fastened with dovetail joints, was used to insulate walls, roofs and even floors. The technique was deployed in a couple of construction projects, the Banque de France in Cholet (Maine-et-Loire) and the municipal theatre in Beauvais (Oise) being two examples of its use. On a wider scale, cardboard has been used more overtly in architecture, producing some interesting results, such as Guy Rottier's corrugated cardboard leisure home, the emergency shelters that Shigeru Ban has been building since 1989 from cardboard tubes made out of recycled paper, and, more recently, Cottrell and Vermeulen's after-school centre in England (see following pages).
Architecture schools have been just as eager to be involved. During the 1980s David Georges Emmerich of the Ecole des Beaux-Arts in Paris carried out research into structural morphology and the use of cardboard for dome dwellings that people would construct themselves.
At the Ecole nationale supérieure d'architecture Paris-Malaquais, Jean Marie Delarue investigates modular assembly and construction methods. And the Ecole nationale supérieure d'architecture Montpellier (ENSAM) has been supporting the work of Thierry Berthomier and Guillaume Bounoure's Waste.arc collective, which is described on the following page.

• *Undulating corrugated structures at Roc Castel / Le Caylar festival, August 2007.*
• *Test models, ENSAM, Easter 2007.*
• *Welcome activities for students at ENSAM, September 2007.*
• *Under the shadow of a folded structure, ENSAM 2006.*

Waste.arc?

"The Waste.arc collective brings together architects and architecture students who are eager to experiment with ways to create pieces of architecture that are more environmentally friendly; the group's name hints towards its goal of making waste useful. In 2005 we created an association at the Ecole nationale supérieure d'architecture in Montpellier that works alongside the collective. The association aims to experiment with the themes of recycling and architecture, as well as raising awareness of them. Its activities consist of building prototypes and producing conceptual frameworks related to subjects such as the countryside, the relationship between architecture and recycling, and notions of inhabitation and temporariness. We hope to follow up the projects that we're currently working on in Languedoc-Roussillon with others in Austria and Sweden (the home countries of two of our members). We're a group of individuals who have been drawn together by a shared ethical outlook and by a desire to make things.

Our work is based on a simple principle that is well understood by anyone who works with space, namely using the negative, making use of empty spaces, using leftover materials, breathing new life into damaged materials, and discovering unexplored artistic and technical fields.

Architecture and recycling

Cardboard has a real image problem. Most people just think of it as a packaging material: disposable, fragile and with a limited life. But in fact it's an effective product that can last for a long time if it's put to the right use and treated correctly. Furthermore, from the point of view of sustainable development and protecting the environment, we need to keep in mind that cardboard comes from a waste material, used paper, that doesn't have many alternative uses. It's important that in future we direct our efforts towards making sensible use of this material in construction projects, because not only is it worth it from an environmental point of view, but it also represents a worthwhile challenge for designers.

We don't make use of all the possibilities that cardboard has to offer, but it could be used as a substitute material, especially in temporary creations, for other more costly and non-renewable resources. Being an architect doesn't have to mean creating permanent structures, nor working with traditional materials.

Folded cardboard

Cardboard comes in to its own when it is folded, as this makes it solid and gives it volume, and also lets you work with shadow and light. Folds, which function both to create form and provide structure, are a particularly interesting feature to us. Our starting point for structures made from folded cardboard is the work of Thierry Berthomier, a lecturer at the architecture school at Montpellier, and, above all, a passionate innovator.

Thierry's work on making use of folds has led to the creation of hundreds of models, each more astonishing than the last. The Waste.arc collective, of which Thierry Berthomier is a member, wants to take Thierry's work into new areas, turning it into a human scale and bringing to life this innovative and poetic architecture that turns the geometric into the organic."

REMARKS BY GUILLAUME BOUNOURE FROM DECEMBER 2007.

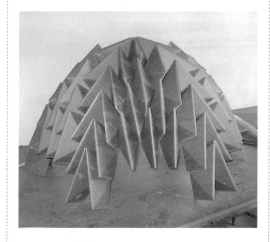

✆ **www.archiwaste.com**

• *Research, Easter 2007.*
• *Prototype models:*
trying out the Boîte de Moulis, *2007.*

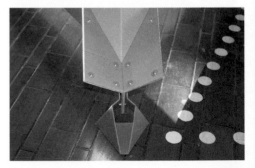

• *Computer-generated image, Chloé Genevaux, 2007.*
• *Exhibiting the architect Shigeru Ban's cardboard footbridge at the Pont du Gard.*
• *Close-ups of the tripods' construction.*

• *Tripods at an ENSAM exhibition.*

Cardboard School

COTTRELL AND VERMEULEN

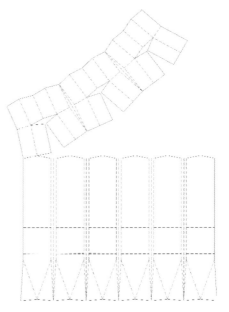

The challenge put to Cottrell and Vermeulen Architects and engineering consultants Buro Happold was to build a structure that made use of the properties of cardboard to create an inspiring play area for after-school activities. In this case, the material itself was the inspiration for the architectural design. The project was made possible by the designers' productive relationship with both the school and cardboard manufacturers and suppliers (Paper Marc Ltd, Essex Tube Windings Ltd, Quinton and Kaines Ltd, and CG Franklin). The potential for building a cardboard structure was examined through experimenting with origami and analysing the structural strength that could be provided through using folds. Aesthetically the building embodies the spirit of origami, which is also expressed through the artist Simon Patterson's artwork of the folding pattern for an origami heron – a species found locally – that decorates the structure. The project achieved its objective of demonstrating the potential uses of cardboard, testing its durability and highlighting a willingness to create a building with a provisional lifespan of twenty years. A large amount (over 90% of the building's materials) of recycled or recyclable cardboard was used in the project. The walls and the roof are made from 166mm-thick composite panels, while the support columns of the structure and the perimeter walls are made from cardboard tubing.

📖 **Westborough Primary School, 2001.**

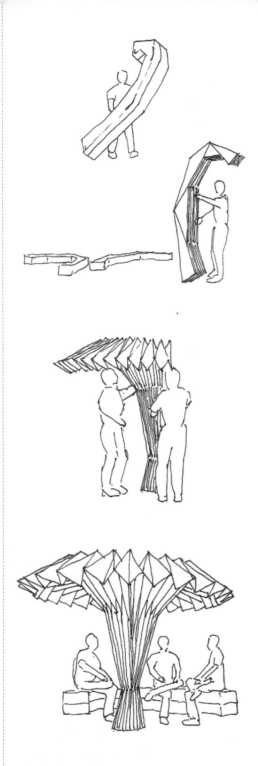

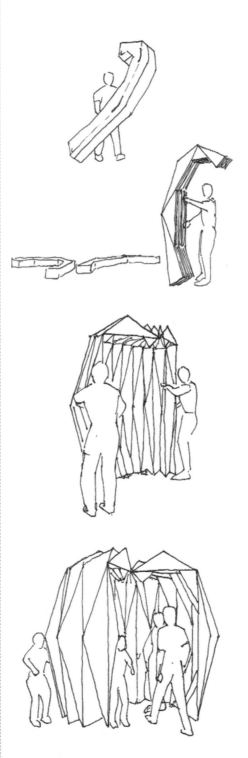

ParaPli

OPHÉLIE BERTOUT AND MARCUS KISTNER

"A pop-up box for the summer getaway" was the theme of the Minimaousse 3 design contest. Fiona Meadows describes the brief: "Who hasn't dreamt, as they take their bathing suit out of their closet as winter comes to an end, of also being able to pull out a feature-laden and surprise-packed summer home that went far beyond an ordinary tent? It would be a sort of adult-scale pop-up which would turn architecture into a game in which a love and enjoyment of construction would be at the forefront, alongside technical and poetical inventiveness. We asked young designers to give shape to this dream by imagining a feather-light and ingenious summer shelter."

ParaPLI, the prize-winning project by Ophélie Bertout and Marcus Kistner of the Ecole Nationale Supérieure d'Architecture in Versailles, takes its inspiration from the poetry of origami: "The piece is designed to be more than just a static structure. The triangle-based folding design allows the structure to be unfolded in two different ways, and therefore into two different shapes. With its dual open and closed forms, the ParaPLI is a fun and lightweight shelter that you can enjoy on your own or with friends."

✆ **www.minimaousse.citechaillot.fr/;**
www.parapli.com

• *Set-up stages of the ParaPLI, illustrating how it can be used in rain, wind or sun. Easy to carry, once it has been taken out of its carry-case it just needs to be unfolded. A hoop is then passed through the loops located at the trunk of the ParaPLI to hold its shape. Once it's been set up, the carry-case can be used as a bench on which to relax or chat in the shade*
• *Right-hand page: the ParaPLI set up in parasol and windbreaker modes.*

Ha-ori Shelter

JOERG STUDENT

After completing a degree in aerospace engineering at the University of Stuttgart, Joerg Student turned his attention to studying industrial design and engineering at the Royal College of Arts in London, where he designed this shelter for his master's dissertation project in 2004. He currently works as a designer at IDEO in Palo Alto. "Creating the structure of the shelter required lots of experimenting with folds, maths and prototypes. The body of the shelter is folded out of a single sheet of 3.5m by 14m corrugated polypropylene. The Ha-Ori is a foldable emergency shelter and as such is designed to be transported, stored and set up quickly and easily. It can be used in both hot and cold environments. Inspired by the structure of hornbeam leaves, the Ha-Ori (which means 'folding leaf' in Japanese) shelter is foldable, light (weighing 36kg in total) and very sturdy, capable of withstanding icy winds and snow. The shelter is made from three parts: the main structure, a central piece with a ventilation system and a door. When folded up it measures 2.6m by 46cm, and can easily be carried by two people and erected in a couple of minutes. When unfolded the shelter provides a space measuring 3.65m across and 2.44m high, enough to accommodate a family. The roofing and the side flaps can be adjusted for ventilation purposes, making it possible to cook or heat the shelter from the inside. Able to wrap around its inhabitants and yet also rigid, the Ha-Ori provides both a physical and emotional refuge. The main structure is made from a translucent, double-skin, high-density, polypropylene sheet, which is both rigid and insulating. The lines of the folds are scored in to the sheet using simple machinery. The properties of the polypropylene allow it to be folded and re-folded an unlimited number of times. The shelter has a potential lifespan of more than 10 years, depending on how it is used. Once it has reached the end of its useful life it can be completely recycled."
REMARKS BY JOERG STUDENT FROM MARCH 2008.

✆ www.cv-arch.co.uk; Hoberman associates; "Habitaflex" house by Laprise; "living box" foldable house;
Flasher by Jeremy Shafer and Chris Palmer; Kobayashi; David Mittchell; Yan Chen; Taketoshi Nojima; Théo Jansen;
http://erikdemaine.org; Tim Tyler: http://pleatedstructures.com ; www.u3p.net;
http://shellhouse.org.
Studio TV "Conque" by Gilles Ebersolt; www.joanmichaels.paque.com; *Nature* journal, July 2007.

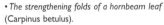

- *The strengthening folds of a hornbeam leaf (Carpinus betulus).*
- *Ha-Ori, folded up.*
- *Erected in Bolinas, CA.*

Crease Patterns

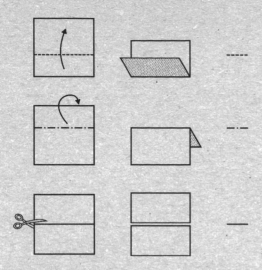

• Origami symbols: valley fold, mountain fold, cut.

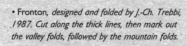
• Fronton, *designed and folded by J.-Ch. Trebbi, 1987. Cut along the thick lines, then mark out the valley folds, followed by the mountain folds.*

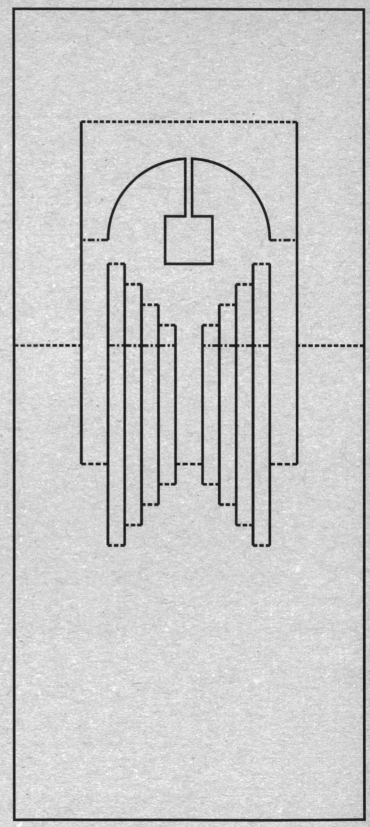

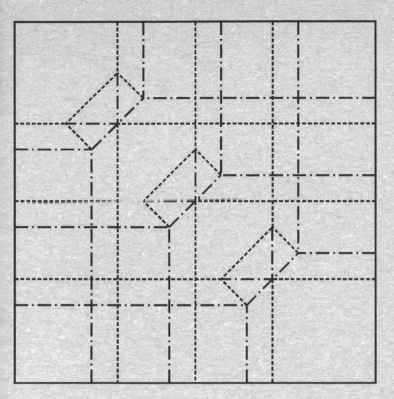

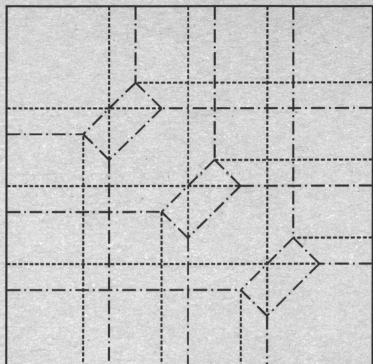

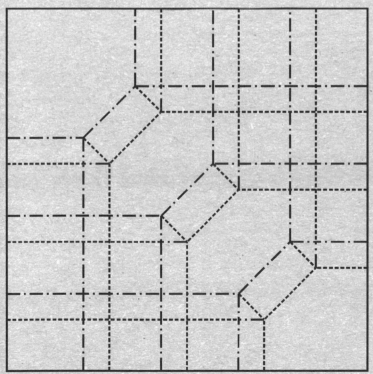

Three patterns inspired by the works of Azuma Hideaki: designed and folded by J.-Ch. Trebbi, 2007. Mark out the folds using a stylus before starting to fold symmetrically as indicated in the patterns and from the following points:
• pattern A, from top-left,
• pattern B, from top-right,
• pattern C, from bottom-left.

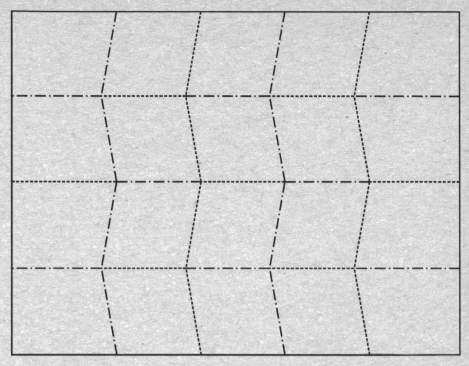

• Miura-Ori *fold, designed by Koryo Miura, 1970.*
• Cape: *designed and folded by J.-Ch. Trebbi, 2007.*

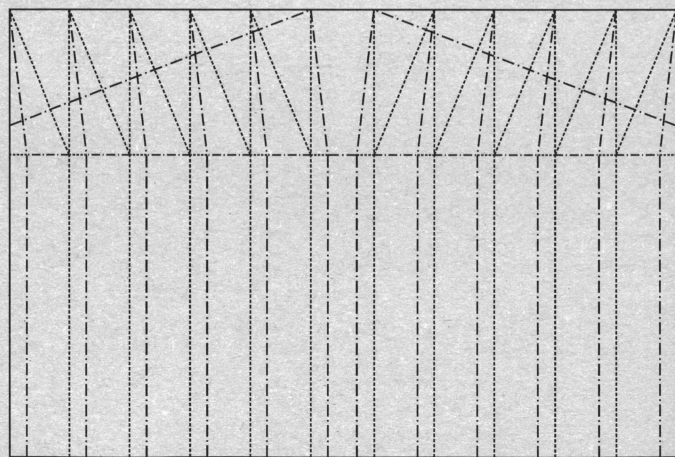

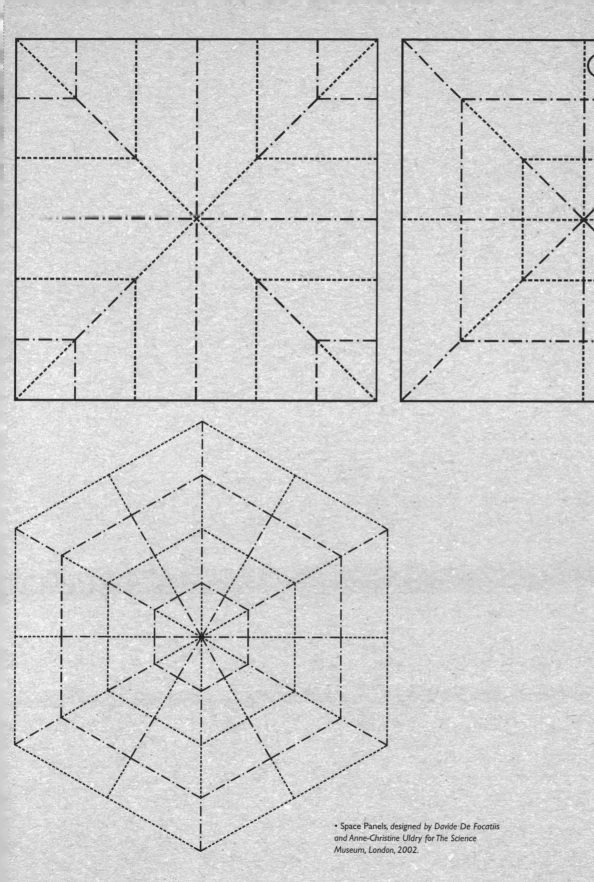

• Deployable structures: designed by
Davide de Focatiis and S.D. Guest, 2002

• Space Panels, designed by Davide De Focatiis
and Anne-Christine Uldry for The Science
Museum, London, 2002.

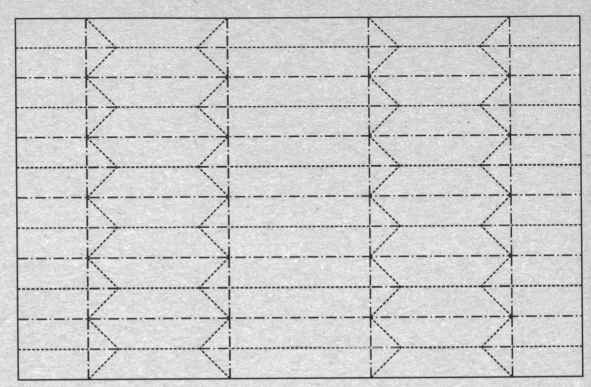

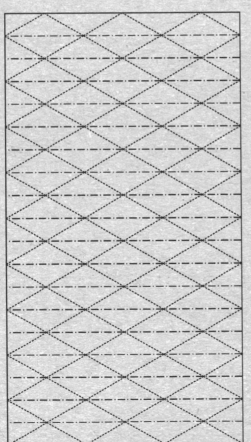

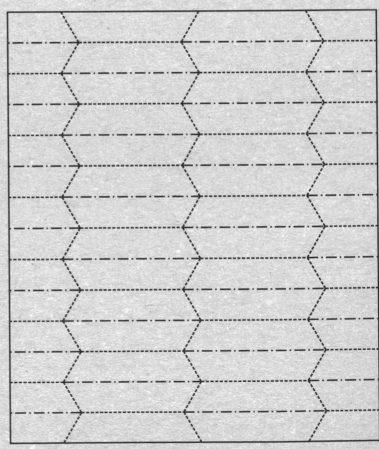

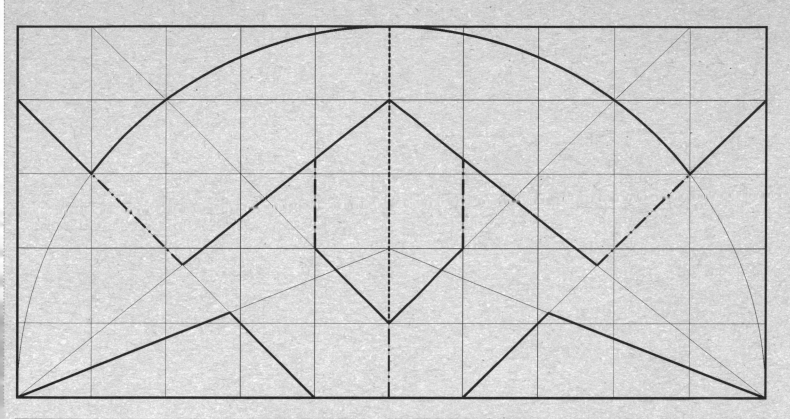

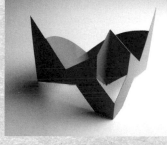

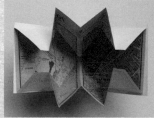

• Gran Balôn, pop-up advertising card,
from the Turin antiques market, Italy.

• Travel sculpture by Bruno Munari.
Mark out the folds, cut along the solid lines and
then make the folds. Model made according to
the plan of the invitation card released on
Munari's 100th birthday by the CLAC Galleria
del Design e dell'Arredamento di Cantù (Como,
Italy).

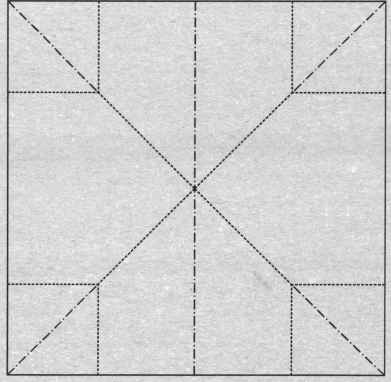

Left-hand page, three accordion folds:
• square: mark out the longest folds and then shape
the triangles
• hexagon: 30 angle, mark out the mountain folds,
then the diagonal valley folds from left to right.
• triangle: 60 angle°, mark out the folds, changing
direction as you go.

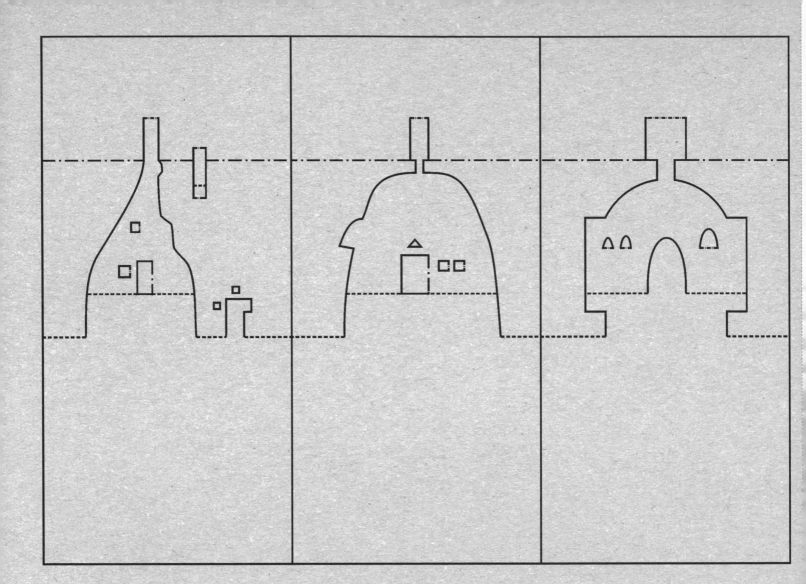

• *Inspired by primitive dwellings:*
Doué-la-Fontaine (France); Santorini (Greece),
and Matera (Italy).
Designed and folded by J.-Ch. Trebbi, 1993.
See Habiter le paysage, ed. Alternatives, 2007.

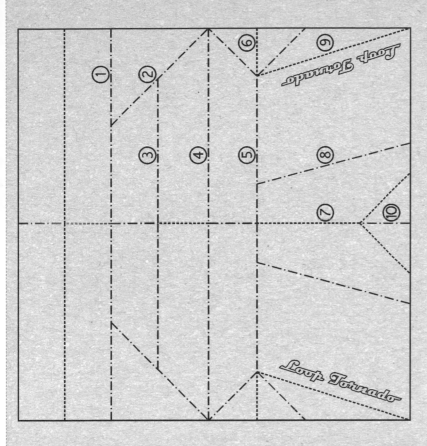

Loop Tornado

Loop Tornado

① ② ⑥ ⑨
③ ④ ⑤ ⑧
⑦ ⑩

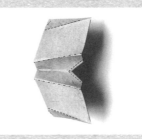

• Loop Tornado *plane adapted from*
Stephen Weiss and Hans R. Bergan's Arc
Wing *by Corentin Delmas, 2008.*

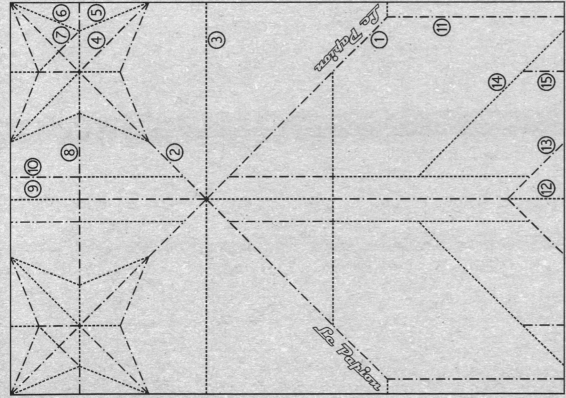

Le Papion

Le Papion

Le Papion

⑥ ⑤
⑦ ④ ③
① ⑪
⑧ ② ⑭ ⑮
⑩ ⑬
⑨ ⑫

• Le Papion *plane, traditional design*
adapted by J.-Ch. Trebbi.

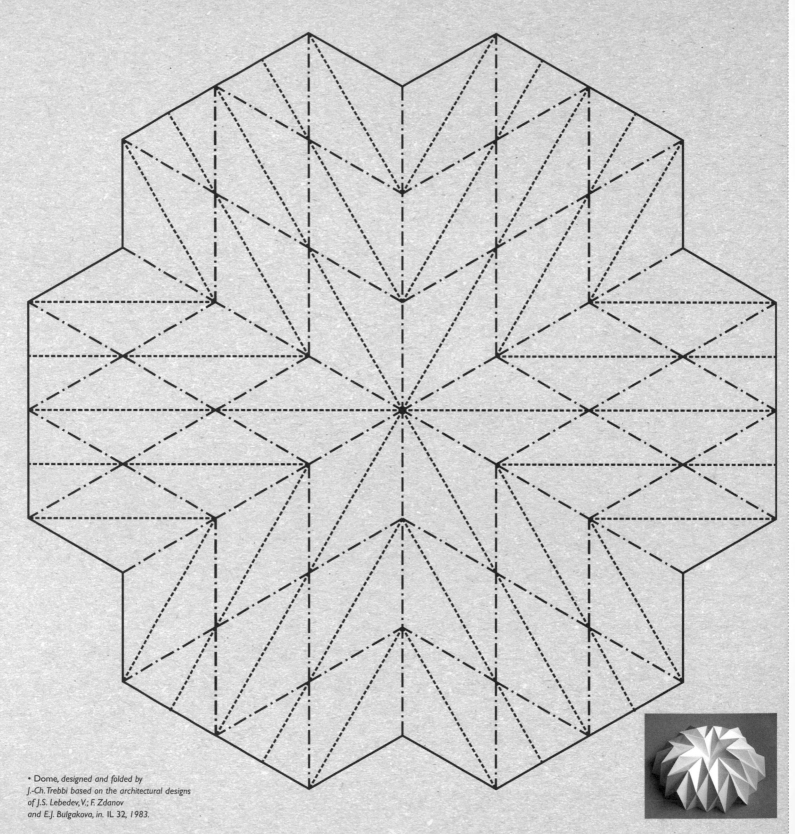

• Dome, designed and folded by
J.-Ch. Trebbi based on the architectural designs
of J.S. Lebedev, V.; F. Zdanov
and E.J. Bulgakova, in. IL 32, 1983.

• Large Dome, *designed and folded by
J.-Ch. Trebbi, based on the architectural
designs of J.S. Lebedev, V.; F. Zdanov
and E.J. Bulgakova, in. IL 32, 1983.*

Acknowledgements

As I look back on our encounters and conversations, my warmest thanks go out to all the designers, architects, artisans, photographers, researchers and creators who have put their trust in me by allowing me to publish their creations, and who have shared their ideas and their passion. I hope that the artists who are not mentioned in this work do not take offence at the omission; the art of folding has turned out to be a limitless world that cannot be contained in a single volume!

Akamine Hidetoshi, Isabella Scistrino (Antonangeli Illuminazione) - Anne - Laure Chevallier (Ecole Pivaut-Nantes) - Annette and Paul Hassenforder (Créations Paula) - Annick Lombardet and Jean-Pierre Campredon (Cantercel) - Benoît Roman, Charlotte Py, José Bico, (PMMH - ESPCI), Charles Baroud - Brian Vermeulen (Cottrell and Vermeulen) - Brig Laugier - Cesare Misserotti and Perla Bianco (Galleria l'Elefante) - Christian Voillot (EFPG) - Christiane Bettens - Christine Phung and Barbara Bouyne - Daniele Lorenzon (Tingo Design Gallery) -Davide de Focatiis and Anne-Christine Uldry (Oxford University) - Diane Steverlynck - Edwin White - Ennio Vicenzoni and Inge de Jager - Enrico Wilde and Mehrwerk Designlabor - Ergian Alberg and Laura Aquili - Eric Singelin - Eva Guillet and Aruna Ratnayake (Studio Lo) - Foreign Office Architects (FOA) and Emilia Izquierdo, Satoru Mishima - Frank Kerdil, Henrik Kirk (Noming) - Frederic Debackere (Borndesign) - Fulvio Ferrari and Napoleone Ferrari (Museo Casa Mollino) - Generoso Parmegiani, Stefania Vairelli (Nomade Design) - Georges Caux, Estelle Martin-Nicolas (Adseat) - Gérard Lognon , Liliane Leboul and their team - Grégory Lacoua, Fabienne Follin, Pascale Frocourd (Ligne Roset) - Guido Daniele, Antonietta Santoli and Laura Fassi (Ad Maiora) - Guillaume Bounoure, Chloé Genevaux, and the WASTE.arc collective - Guy Rottier and Véronique Willemin - Hannah Allijn and Egwin Heins - Heatherwick Studio, Catherine Kenyon and David Balhuizen - Helmina Sladek and Pauline Thierry - Helmut Frerick (La Font du Ciel) - Hiroki Takada (Takada Design) - Inga Sempé - Isamu Tokizono (Brave Design) - Jean-Louis Pinabel, Geneviève Baccouche and Jean-Marc Germain - Jean-Marie Delarue - France and Jean-Paul Moscovino - Jean-Yves and Dominique Doyard (Moulin de Kéréon) - Jérémy Bataillou, Michel Pagliosa, Laure Marty

(Atelier du Béton) - Joan Sallas, Maria Richter, Birger Peil, Markus Schwellensattl, Ulrich Ghezzi - Joerg Student, Sue Bradburn (Royal College of Art, Kensington Gore), Anna Weldon (Palo Alto Art Center) - Josh Jakus - Julien Gritte (Irigimi) - Jun Maekawa - Kouichi Okamoto and Ayako Nakanishi (Kyouei design) - Laura Aquili and Ergian Alberg (AquiliAlberg) - Laurence Millet - Lisamarie Villegas Ambia and Philippe Dufour Feronce - Luisa Canovi (Paper Factory) - Manuelle Gautrand and Mélanie Trinkwell - Margriet Foolen - Marie-Christine Renard (Plissés de France) - Marie Compagnon, Guillaume Hillairet , Filloux & Filloux , Cécile and Stefano Poli - Marion Bataille - Masahiro Chatani - Matthias Ries - Max Hulme - Michel Charbonnier - Nicola Enrico Stäubli - Olivier Boisard (U3P) - Olivier Draullette (Ondef) - Olivier Leblois, Isabelle Millet , Stéphane Millet and Audrey Batacchi (Quart De Poil) - Ophélie Bertout and Marcus Kistner - Paul - Henri Jeannel (Chapeau Magique) - Peter Karpf, Gunar Ottosson (Iform-Voxia) - Philippe Huger - Pietro Seminelli and Kathy Vaubaillon - Raphaël Passas (EFPG) - Raquel Vega and Vanessa Cardoso Santos - René Lachèze and Hélène Migot (Maugein accordéons) - Santiago Calatrava and Sonia Elisario - Sonia Biacchi, Carla Marazzato (CTR) and Kristine Thiemann, Giorgio Bombieri - Akiko Miyaké - Stefan Delecat (Hannecke) - Stéphan Diez, Christophe de la Fontaine and Dominik Hammer - Sumiko and Yoshihide Momotani - Sylvain Le Guen (LG éventails), Serge Davoudian (Le Curieux) and Thierry Jacob - Sylvie Dameron - Taki Girard - Tamami Kojima (Mito Arts Foundation) - Thomas Dumoulin and Nicolas Marzouanlian - Timm Herok (Foldtex) - Tyne De Ruysser - Valentina Fortuna and Raffaella Zuccarini (FIAM) - Valérie Jomini and Stanislas Zimmermann - Véronique Wardega (Petits Plis) and Claire Lascaux - Wilm Fuchs and Kai Funke (Industrial Design) -Yoshiharu Tsukamoto and Yoshiko Iwasaki (Bow Wow) - Zhong You and Kaori Kuribayashi (Oxford University).

I offer my warmest gratitude to the following people for their assistance and collaboration: Annette and Patrick Castel, Johanna and Genghis -

Agostina Pinon (IFA) - Alessio (Edizioni Corraini) - Charles Shopsin (blog.modernmechanix) - Christiane and Gaby Lemaire - Corentin, Marie-Christine and Christian Delmas - Danièle Angevin (Escarboucle) - Eléonore Durand (Pôle Verrier-Nancy) - Elsa Bigey - Eric Lafuma (PL- Diffusion) - Eva Hartmannsgruber (Kerafol) - Gérard Laizé (VIA) - Jacques Desse and Thibaut Brunessaux (Boutique du livre animé), Thierry Desnoues - Jean-Jérome Casalonga - Lionel Turban (Disactis) - Liliane Guizzoni - Madame Crosio (Fratelli Crosio Accordéons) - Mark Pollard (Form & Frys) - Martine Cebron - Martine Tripet - Michel Mortier (Arcelormittal) - Milena Sosa - Mireille Mazet - Myriam Feuchot (IFA, Cité de l'architecture & du patrimoine) - Patrick Bertholon - Patrick Bouchain - Patrizia Ghirardelli and Massimo Missiroli - http://www.philcad - Philippe Potié - Regina Kaltenbrunner (Musée Baroque de Salsburg) - Richard Combes - Serge Tisseron - Siria Rizzi and Mieke Claes (Candidus Prugger) - Toda Takuo and Yagishita Makoto (Castem Inc.) - Yuki Otaka (Association Avion Origami - Paris).

Looking back on the encounters I had in Paris in May 2008, I would like to thank the members of the MFPP, and in particular : Alain Georgeot, Aurèle Duda, François Dulac, Louis Eveillard, Naomiki Sato, Olivier Viet, Shang Lee, Tsan-Liang Lieu, Véronique Lévêque, and Yves Clavel .

I am grateful to all the team at Editions Alternatives for the pleasurable experience of creating this book, and would like to thank: Catherine Paradis, Charlotte Gallimard, Deborah Van Quickenborne, Gérard Aimé, Patrice Aoust and Sabine Bledniak, as well as Denis Couchaux for his invaluable collaboration and his superb graphic design work.

I would like to extend my personal thanks to Claudine Pisasale, a masterly origamist and modeller, for the energy and passion with which she conveys her knowledge of the art of folding; and to Nicole Charneau-Trebbi and Denis Trebbi for their encouragement, patience, understanding, support and day-to-day advice, for having accompanied me along the way as this adventure has unfolded, and for having stretched themselves not just to translating and designing graphics, but also to accepting the burdensome work of fold testing!

Bibliography

Architectures à découper, exhibition catalogue, Arc en Rêve, Bordeaux, 1987.
ALBERTINO Lionel, Origamania.
BADALUCCO Laura, Kirigami, Celiv, Paris, 1997.
BOURSIN Didier, Avions de papier en origami, Dessain et Tolra, Paris, 2004.
BUISSON Dominique, Japon papier, Pierre Terrail, Paris,1991.
CHATANI Masahiro, Origami architecture, American houses, Kodansha International, 1988.
CHATANI Masahiro, Origamic Architecture of Masahiro Chatani, Shokokusha, Tokyo, 1983.
CHATANI Masahiro, Pattern Sheets of Origamic Architecture, Shokokusha, Tokyo, 1984.
CHATANI Masahiro, Four Seasons of Origamic Architecture, Shokokusha, Tokyo, 1984.
CHATANI Masahiro, Key to Origamic Architecture, Shokokusha, Tokyo, 1985.
CHATANI Masahiro, Origamic Architecture Around the World, Shokokusha, Tokyo, 1987.
DELARUE Jean-Marie, Morphologie, UPA1, Paris, 1978.
DELARUE Jean-Marie, Faltstrukturen, in IL N°27, Natural Structures, 1980.
DELARUE Jean-Marie, Pliage, Institut de recherche en morphologie structurale, Paris, 1980.
DELARUE Jean-Marie, Plis, Règles géométriques et principes structurants,
Ecole d'Architecture Paris-Villemin, Paris, 1992.
DELARUE Jean-Marie, Morphogenèse, Paris-Villemin, Paris, 1992.
ENGEL Peter, Folding the universe, Random house, New York, 1989.
IL 32, Lightweight Structures in Architecture and Nature, Moscow, 1983.
JAKSON Paul, Pliages et découpages, Manise, Paris, 1996.
KASAHARA Kunihiko, Origami, Nichibou, Japan, 2004.
KENNEWAY Eric, Complete origami, Ebury Press, London, 1987.
Les cahiers de la recherche architecturale, Imaginaire technique, Parenthèses, 1997.
MICHEL-DUBRETON M.J., Pliez du papier, Fleurus, Paris, 1973.
MIURA Koryo, Origami science et art, Otsu, Japan, 1994.
MOMOTANI Yoshihide, Incised origami, Seibundo, Japan, 2001.
PAIREAU Françoise, Papiers japonais, Adam Biro, Paris, 1991.
PAULAIS Martine, Papier, créations et métamorphoses, Dessain et Tolra, Paris, 2006.
PARRAMON J. M, PERIS J. J., Travaux manuels en papier et en carton, Bordas, Paris, 1977.
RÖTTGER E., KLANTE D., Le Papier, le jeu qui crée, Dessain et Tolra, Paris, 1969.
SCHÜSSLER Brigitte, Papierarchitectuur, Cantecleer, de Bilt, 1990.
TISSERON Serge, Petites mythologies d'aujourd'hui, Aubier, 2000.
THACKERAY Beata, L'Art du Papier, Gründ,1997.
VIET Olivier, Avions en origami, Fleurus, 2004.
VYZOVITI Sophia, Folding Architecture, BIS Publishers, Amsterdam, 2003.
VYZOVITI Sophia, Supersurfaces, BIS Publishers, Amsterdam, 2006.
WARDEGA Véronique, Bijoux en origami, L'Inédite, 2006.
WEISS Stephen, Wings and Things, Origami That Flies, St. Martin's Press, New York, 1984.
ZÜLAL AYTÜRE - SCHEELE, Origami, Fleurus idée, Paris, 1997.

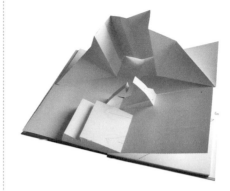

• Trente-six, *designed and folded by Eric Singelin, June 2008.*
• A la courte échelle, *Philippe Huger, 2008.*
• *Left-hand page: Christine Phung creation, pleating by Atelier Lognon, 2008.*

Picture Credits

Cover: Pietro Seminelli; p. 4 tl: France Moscovino; p. 4 tm: Joerg Student; br: Kristine Thiemann; bl: Patrick Castel; p. 5 tl: Quart De Poil; tr: Museo Casa Mollino; bl: Waste. Arc; p.6 bl: Claudine Pisasale; p. 8 tl: Jean-Jérome Casalonga; p. 8 bl: Denis Trebbi; p. 10 t: Guido Daniele; p. 11 tr: Toda Takuo; p. 12 tl: Mark Pollard; p. 12 bm: Escarboucle; p. 14 tm Maugein Frères; mc: Jean-Jérome Casalonga; p. 14 bm: Casa Mollino; p. 15 Casa Mollino; p. 16 tl: Lionel Turban; p. 16 bm: Charles Shopsin; p. 16 bl: Henry Krul; p. 17 mc: Zhong You; p. 17 mc: Benoît Roman, Charlotte Py, José Bico; p. 18: Kristine Thiemann; p. 19 tr: Christiane Bettens; p. 19 mr and br: Kristine Thiemann p. 21 bl and br: Satoru Mishima; p. 22: Kouichi Okamoto; p. 24 tl: Taki Girard; p. 25 ml: Claudine Pisasale; p. 25 tr and bm: Denis Trebbi; p. 26 ml: Raphaël Passas; p. 26 tr and 27: Dominique Doyard; p. 30 tl: Nicole Charneau-Trebbi; p. 32 t, b and m: Luisa Canovi; p. 34 bl: Cécile and Stefano Poli; p. 34 tl and p. 35 tr: Guillaume Hillairet; p. 34 tm and p. 35 bl: Marie Compagnon; p. 36 tl: Galleria Elefante; p. 37 tl: CLAC and munart. org private collection; p. 37 br: Luciano Olzer; p. 38 ml, mr, br and p. 39: Jean-Paul and France Moscovino; p. 40 Seth Tice-Lewis; p. 41 t det b: Edwin White; p. 41m: Dennis Kok; p. 42 tl, bl, and bm. Artowermito; p. 42 br and p. 43: David Balhuizen; p. 44 tl: Le Klint; p. 44 br and p. 45 tl: Kouichi Okamoto; p. 45 tr: Inga Sempé; p. 45 bl and ml: Isamu Tokizono; p. 46: Michel Charbonnier; p. 47: Helmut Frerick; p. 48 and p. 49 bl: Denis Trebbi; p. 49 mc and r: Antonangeli Illuminazione; p. 50 t: Padore; p. 50 br: Ulrich Ghezzi; p. 51 bl and br: Birger Peil; p. 51 tr: Markus Schwellensattl; p. 52 t: Diane Steverlynck; p. 50 b and 51 tl: PL- Diffusion; p. 53 bm and r: Noming; p. 54 and p. 55 : Kérafol; p. 56 bl, p. 56-57 m, and 59 tl: Thierry Jacob; p. 58 and p. 59 br: Sylvain Le Guen; p. 60 tl and br: Tyne De Ruysser; p. 60 mc, br and p. 61: Claire Lascaux; p. 62: Studio Lo; p. 63 tr and mr: Josh Jakus; p. 63 br: Sylvie Dameron; p. 65: Créations Paula; p. 66 and 67: Jean-Marc Germain; p. 68, 69, 70 , 71: Denis Trebbi; p. 72, 73 and 74: Pietro Seminelli; p. 75: Egwin Heins; p. 76: Takada Design; p. 79: Candidus Prugger; p. 80: S. Cavallo; p. 81: Museo Casa Mollino; p. 82 and 83 b: Voxia; p. 83 t: Margriet Foolen; p. 84 d and 85 : Foldtex ; p. 86 and 87 : Anne - Laure Chevallier ; p. 88 and 89 : Quart De Poil ; p. 90 and p. 91 tl and ml: Nomade Design; p. 91 tm and r: Valérie Jomini and Stanislas Zimmermann; p. 92: Industrial Design, Fuch and Funke; p. 93 tl: Nicola Enrico Stäubli; p. 93 r: Adseat; p. 94: Borndesign; p. 95: Mehrwerk Designlabor; p. 9 : Thomas Dumoulin and Nicolas Marzouanlian; p. 97: Matthias Ries; p. 98 and 99: Vanessa Cardoso Santos; p. 100: Stéphan Diez and Christophe de la Fontaine; p. 101: Ennio Vicenzoni and Inge de Jager; p. 102 and 103: graphics by Lisamarie Villegas Ambia and Philippe Dufour Feronce; p. 104: Ligne Roset; p. 105: Takada Design; p. 106 and 107: Cut'nfold, Hannecke; p. 108: AquiliAlberg; p. 109 : Fiam Italia; p. 110 and p. 111: Atelier du béton; p. 112 ml: Marie Compagnon; p. 112 tr and p. 113 tl, br and tr: Guillaume Hillairet; p.111 bl: Filloux & Filloux; p. 114 and 115: graphics by Helmina Sladek and Pauline Thierry; p. 116 tl and ml: Sim Van der Ryn; p.116 cl: J.M Prada; p 116 bl: Bow Wow Architects; p 116 br: Denis Couchaux; p. 118 tl, br and 119: Guy Rottier; p. 120, 121, 122 bl and br and 123 tl: Waste. Arc; p 122 tl: graphics by Chloé Genevaux; p. 123: Brian Vermeulen; p. 124 and 125: graphics by Ophélie Bertout and Marcus Kistner; p. 126 and 127: Joerg Student; p. 128: Christiane Bettens; p. 140 ml: Nicole Charneau-Trebbi; p. 141 tr: Eric Singelin; p. 142 tl: Denis Couchaux; p. 142 bl: Claire Lascaux; p. 9, 13, 129 to 139: graphics of diagrams and folding patterns: Denis Trebbi.

Photos not listed here are by the author.

In spite of our best efforts we have been unable to identify the photographers of some images; we are prepared to make any necessary changes once these photographers' copyright over their images has been authenticated.

THE ART OF FOLDING: CREATIVE FORMS IN DESIGN AND ARCHITECTURE
ORIGINAL TITLE: L'ART DU PLI: DESIGN ET DÉCORATION
TRANSLATOR: TOM CORKETT. LAYOUT AND GRAPHIC DESIGN: DENIS COUCHAUX. COVER DESIGN: GREGORI SAAVEDRA
ISBN-13: 978-84-92810-66-6

© EDITIONS ALTERNATIVES FRANCE, 2008, FOR THE ORIGINAL VERSION
© PROMOPRESS 2012 FOR THE ENGLISH-LANGUAGE EDITION
PROMOPRESS IS A COMMERCIAL BRAND OF: PROMOTORA DE PRENSA INTERNACIONAL S.A., C/ AUSIÀS MARCH 124, 08013 BARCELONA, SPAIN TEL.: +34 93 245 14 64 FAX: +34 93 265 48 83 EMAIL: INFO@PROMOPRESS.ES
WWW.PROMOPRESS.ES WWW.PROMOPRESSEDITIONS.COM
FIRST PUBLISHED IN ENGLISH: 2012

PRINTED IN CHINA

• *Zeiss-Ikon bellows camera.*
• *Origami by Véronique Wardéga.*

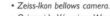